WARBIRDS WORLDWIDE

Volume Three Number Four

Founders:
Paul A. Coggan
John R. Sandberg
Henry J. Schroeder III

Director and Editor:
Paul A Coggan

Financial Adviser
Philip S. Warner F.C.A.

Staff Photographer
Thierry Thomassin

Air Racing Correspondent
Sharon K. Sandberg

Chief Pilot:
Robb R. Satterfield
(Lt. Col. USAF Ret'd)

Photographers:
Peter Anderson (Australia)
Robert S. DeGroat (USA)
Jeremy K. Flack (UK)
Jack Flinn (USA)
Alan Gruening (USA)
Robert Livingstone (Australia)
Philip Makanna (USA)
Ron Miller (USA)
Frank Mormillo (USA)
John Rigby (UK)
Michael J. Shreeve (UK)
Chuck Sloat (Canada)
Nick Veronico (USA)
Philip Wallick (USA)

Editorial Address:
Five White Hart Chambers
16 White Hart Street
Mansfield
NOTTS. NG18 1DG
England
Tel: (0623) 24288
Fax: (0623) 22659

WARBIRDS WORLDWIDE LIMITED
Company No. 2107572
Registered Office:
Winchcombe House
Winchcombe Street
Cheltenham
GLOS GL52 2NA

CONTENTS

The Champlin Fighter Museum Fw190D - see Page 14 for the full story

4 OPINION
The Editor looks at the last three years of Warbirds Worldwide, its achievements, and the warbird scene in general

6 MUSEUM OF FLYING AUCTION
We reveal plans for what must be the aviation event of the decade in Santa Monica, California next May.

14 CHAMPLIN Fw190
Alan Gruening details the restoration to engine running condition of the prize of the *Champlin Fighter Museum* collection.

16 SEA FURY
Graham Trant outlines the current Sea Fury scene and some exciting plans for the future.

18 FLYING *SUGARLAND EXPRESS*
Anders K. Saether tells of his experiences whilst flying the SHF A-26B Invader.

29 THE TERRIFIC TROJAN
Rolf Meum takes us on a trip in the SHF North American T-28C Trojan.

33 CLASSIC AIR SERVICES
Jeffrey Ethell travels to Cape May, New Jersey, to view a recent newcomer to the commercial warbird restoration scene.

36 THE AWESOME MiG-21
Finnish Air Force test pilot Jyrki Laukkanen takes us on a high speed trip in the newest of warbirds, the *MiG-21*

- **SCENE P9**
- **JETTOPPICS P36**

WARBIRDS WORLDWIDE is published in book form every three months. It is the official publication of the organisation of the same name and acts as a permanent published record and a forum for the exchange of information and ideas amongst our membership. It is provided free of charge to members and is available at the cover price to non-members. Details of membership fees are available from the Editorial address: Membership is open to anyone with an interest in warbirds and warbird ownership is not a qualification required to join.

What a magnificent achievement! Nobuo Harada's Mitsubishi Zero Sen Zeke in Tokyo, Japan. Nubuo is still looking for a suitable powerplant. This is an indication of the geographic spread of the warbird movement!

OPINION

The Editor looks at the first three years of *Warbirds Worldwide*, our achievements and future plans.

When Butch Schroeder, John Sandberg and I formed *Warbirds Worldwide* some three years ago we knew the warbird movement was about to see an upsurge in activity. However, we had no idea of the magnitude of the increase. Not only have we seen the anticipated increase but have witnessed a larger geographical spread of activity and a blossoming of the warbird jet movement that no one could have foreseen.

Warbirds Worldwide had humble beginnings under the name of a smaller, but no less dedicated organisation. We started with photocopied sheets stapled in one corner, produced at 'the office' and stapled together in my lounge. It grew to a black and white 24 page newsletter with an even more dedicated following of enthusiasts and owners. At the time I was following a career in the Royal Air Force. I was almost obsessed with aviation, and now, almost twenty years later, I reflect upon those days with fond memories.

Not only are we entering a new decade, but a new era where Eastern Europe is breaking free of its restrictive governments. This is an exciting time. We have already seen a steady exodus of equipment from several Warsaw Pact countries - mainly jet equipment which has been exported to the UK, USA and Australia. Australia! Who would have dreamt we would ever see MiGs flying there? But we will, and all because a few dedicated individuals chose to exercise their right to fight to preserve aviation heritage. The State Department in the USA will hopefully soon be more tolerant of similar imports, once they see the calibre of person they are dealing with - professionals striving to preserve aviation heritage for future generations. It may sound romantic, but this is a serious business. The jet movement has really moved apace. We are constantly reminded about the ice cream parlour and the F-86. That is in the past! Forget about it! We have a new generation of more serious, professional jet rebuilders and operators who are spending vast amounts of money and expending a large amount of valuable time to ensure that the first generation of jets is preserved; just as our predecessors did with the piston engined warbirds. As this is being prepared for printing I will be in the United States attending the *Classic Jet Aircraft Association's* Annual Convention at Tucson, Arizona. Here, this growing association will discuss plans to regulate the jet movement, to liaise with each other to improve flight safety at all levels of jet operation and to talk about training programmes. They are professional people, some with military backgrounds and wide experience of jet operations. When the Federal Aviation Administration stepped in and tried to impose regulations without reference to the operators, the *CJAA* interjected and insisted, in the nicest possible way of course, that the subject should be discussed. And it has been - more in the next issue. Suffice to say the jet movement is here to stay. I am pleased to say that in general we have also been privileged to witness an improvement in rebuilding and restoration standards. In more recent years we have seen Governmental departments in the UK dealing more and more with private individuals. Exchange deals are becoming more common, much to the consternation of some it would seem. Our

Bob DeGroat caught this anonymous P-51D Mustang flying in the United States recently. 1990 sees the 50th Anniversary of this most famous of American fighter aircraft.

policy on this is clear. We support it, and so do the vast majority of our members, for such moves will undoubtedly lead to more airworthy warbirds. We have also seen some notable exchanges of warbirds between the UK and the USA, and again we support such movements. We support the legal, balanced exchange of warbirds, and see no point in exploding into sensationalism every time a Spitfire is exported.

But what, I hear you ask, have you achieved with *Warbirds Worldwide*? In just three short years we have built a stable, truly worldwide membership with members in some 30 different countries. We have taken the lead in covering the expansion of the warbird jet movement. We have produced some sixteen high quality publications, some of which are already out of print. We have introduced engineering companies into the warbird field with great success, creating jobs and saving warbird operators vast sums of money. We have introduced prospective employers to individuals and vice versa, finding employment for many of our skilled members wanting to join the growing band of commercial warbird rebuilding operations. We have introduced operators of similar types to each other with great success. Above all, we have created an outstanding network of contacts, all feeding information into our forum for the exchange of information and ideas; the journal, which in itself has become a collectors item. Quality both in terms of production and information, banishing rumour and idle gossip to other, lesser publications who achieve nothing and do our movement harm with their inaccurate reports. We intend to continue to lead the field, and now that we are at the end of the third volume of the *Warbirds Worldwide* Journal we will be undergoing a design change with effect from the next edition, but maintaining the easy to read format and quality of production and information. Producing each journal is a hard slog. Our contributors should share some of the credit for the success of *Warbirds Worldwide*. But most of all it is you, the members and readers that should be congratulated for supporting us. And enabling us to succeed. On the Sundays at airshows when all sane people are relaxing over a leisurely lunch you are queueing in traffic waiting to get into the airfield to see the show. Or wrapping up against that breeze that always howls across the airfield at Duxford. Or dashing in between the squally showers. Shouting above the roar of the Merlin engines and picking the grass (green variety) out of the coffee you have just poured. We are always pleased to see you, whatever country we are in, and hope to enjoy your support for many, many years to come. We are, more than ever before, committed to our cause, and will continue to back those that fight to preserve our warbird heritage by keeping warbirds airworthy. **WW Paul A. Coggan**

Do you have any comments on the content or style of *Warbirds Worldwide*? Whilst we cannot promise to answer each letter individually please write in and outline your ideas. We would certainly appreciate constructive criticism and look forward to hearing from you. Write to Opinion, Warbirds Worldwide, Five White Hart Chambers, Sixteen White Hart Street, Mansfield, Notts NG18 1DG, ENGLAND.

The Santa Monica based Museum of Flying *have an impressive line up of aircraft for their auction next May including the aircraft shown here. North American Mustang racer* Dago Red, *the ex-John Silberman P-38 Lightning, the ex-Larry Barnett Spitfire from South Africa and the P-40 are all much sought after and we could well see some hefty bidding.* (Museum of Flying Photo)

Museum of Flying Auction

Paul Coggan examines plans for a most ambitious aviation auction at Santa Monica in California. Not only is there an impressive line up of lots, there is a host of dynamic aviation personalities promoting the event

Billed as 'the largest auction of pre-1948 aircraft ever held' by the organisers, the Santa Monica based *Museum of Flying*, this incredible event is scheduled to take place on Saturday and Sunday May 19th and 20th at the Museum's own immaculate facility. No doubt this unprecedented event will be of compelling interest to the growing number of warbird and vintage aircraft collectors and investors all over the world.

Several Classic warbird fighters are being offered for sale including Lockheed P-38L Lightning N5596V/44-26981, which was recently sold to the Museum by John Silberman, North American Mustang N5410V/44-74996 (the Reno racer *Dago Red*), Spitfire HF Mk IXe N930LB/MA793 thought to be the only surviving Spitfire that saw service with the USAAF, and a P-40K Kittyhawk N293FR/42-9749.

In addition to the aircraft listed above some 300 pre-1948 aircraft are being evaluated for acceptance at the time the press release was issued from the Museum Auction Office. The organisers stated that they expected to be processing applications for at least a thousand aircraft before the cut off date. "Only prime examples of each class will be accepted for inclusion in the auction inventory..' John Hanley told the *Warbirds Worldwide* office. In addition to the impressive line up of aircraft a substantial number of museum quality items from the Donald Douglas collection including scale models of experimental aircraft, wind tunnel models, cargo aircraft, missiles, space stations and dioramas will also be on sale. Even *Rosie the Riveter's* actual tool kit will be up for auction!

Noted actor (and Spitfire owner) Cliff Robertson is the celebrity spokesperson for the event and will serve on the Auction Advisory Board with some of the aviation industry's most dynamic personalities including Donald Douglas Jr., test pilots Robert 'Bob' Hoover, Scott Crossfield and Tony Levier, former *Federal Aviation Authority* Chief Donald Engen, *Voyager* pilots Dick Rutan and Jeanna Yeager, Paul MacCready, pioneer designer of Human powered aircraft, historian Walter J. Boyne, Moya Lear, widow of famed aircraft entrepreneur Bill Lear, Clay Lacy, Captain of the legendary 747 round-the-world Freedom Flight, motion picture producer Tony Bill and *Mercury, Gemini* and *Apollo* Astronauts Walter Schirra and Buzz Aldrin. The whole event reads like a *Who's Who* of the aviation fraternity.

The organisers have obviously invested a lot of capital in the run up to the auction including some impressive media packages. Proceeds raised from the auction, say the organisers.. "will provide an ongoing endowment for the *Museum of Flying* as well as scholarships for young people to pursue careers in the aviation industry"

This will undoubtedly be the aviation auction of the decade, unparalleled for contents – particularly the warbird section, which sees not just one, but four sought after aircraft. The auction is bound to attract not only warbird collectors but those interested in warbirds purely for investment, and this may see some high bids being submitted, and

CONTINUED ON PAGE 8

Museum of Flying Auction
Continued from Page 6

aircraft being sold, subject of course to realistic reserves being placed on them. *Warbirds Worldwide* will monitor events with interest and report in the next issue on the outcome of what may be the first of many auctions from one of California's primary Museum collections.

WW Paul A. Coggan

One of the stars at the Museum of Flying Auction at Santa Monica is this P-40K 42-9749 / N293FR.

SEA FURY

One of the Sea Furies that does not get a lot of attention is Ellsworth Getchell's Sea Fury FB.11 WH587, construction number 41H/636334, which was bought on charge with the Royal Australian Navy in March 1952. The aircraft has seen a succession of civilian owners including James Fugate, Sherman Cooper, and Westernair of Alberquerque. (Dick Philips)

WARBIRDS WORLDWIDE SCENE

Australia

Several reports have come in recently from Peter Anderson. The biggest news on the Australian warbird scene has been the first flight of **Col Pay's P-40 – Kittyhawk 1A** VH-KTH/AK762 on 15th December 1989. This is the first P-40 to fly in Australia since the end of World War II.

Further P-40 news concerns the **recovery of two rare P-40F Warhawks** from the South Pacific and their planned airworthy restoration in Australia. These aircraft were the survivors of a flight of four aircraft which force landed during the war and were subsequently coverted to components and then abandoned. Although well known for many years, there had been no concerted efforts to obtain the airframes and their acquisition by an Australian group is considered a real coup. The P-40F is the Merlin engined variant and the new engine did much to overcome the shortcomings of the Allison powered machines. The new powerplant was also responsible for name change to Warhawk. There will be full reports on the restoration of Col Pays Kittyhawk and the recovery of the P-40Fs in the next edition of Warbirds Worldwide.

Australian Warbirds held its annual formal dinner and presentation night in Canberra, A.C.T., on the weekend of October 1st/2nd 1989. In addition to warbird owners, represetives also attended from the Civil Aviation Authority and the Federal Airports Commission. The benefits of this face to face exchange between operators and the C.A.A were obvious following the dinner but even more tangible was the waiver of all airport charges granted by the FAC for any warbirds that attended the meeting.

The 1989 Grand Champion Australian Warbird Award was presented to Guido and Lynette Zuccoli for the magnificent Fiat G.59B now registered VH-LIX. Built by Sanders Aircraft Inc. of Chino, California the Fiat is almost unique. The 1989 warbird achievement award was presented to Bob De la Hunty and Gordon Glynn for their efforts in saving, restoring and ferrying the *ex Aeronavale* **P2V-7 Neptune** from Tahiti to Australia (more on this type in future editions of *Warbirds Worldwide* – Ed.)

The jet scene also continues to blossom in Australia with the import of several **BAC Strikemasters** (see Jettopics).

Classic Aviation of Bankstown has recently acquired the badly damaged **Fairey Firefly AS.6 WD828/VH-HMW** together with its complete stock of engine and airframe spares. This aircraft had suffered an engine failure during a test flight from Camden airport, NSW, on 5th December 1987 which resulted in a forced landing that broke the aircraft's back (See *Warbirds Worldwide* Number Four, Page 8). At the time of the accident it was believed the aircraft would never fly again, but following investigations by *Classic Aviation* a

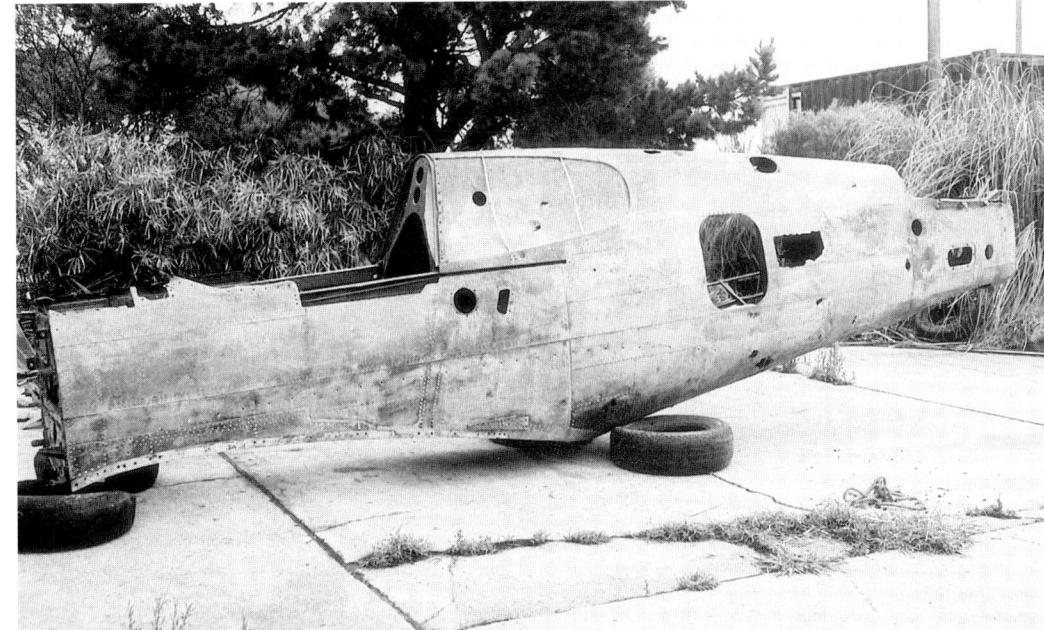

Top: *One of two rare P-40F Warhawks recovered from the South Pacific* (Peter Anderson).
Lower: *At Duxford* The Fighter Collection *are working on this Hawker Tempest for the RAF Museum* (Michael Shreeve)

long term rebuild is considered possible.

The airframe was transported from Nowra Naval Air Station to Bankstown in February last year and a detailed damage survey is currently being undertaken. At the completion of this survey the airframe will be totally disassembled and the rebuild will commence on a part time

WARBIRDS WORLDWIDE SCENE

basis. Due to the substantial damage sustained during the accident no time limit has been set for the rebuild and major airframe components are currently being sought to aid the project. Anyone who can help with either parts or information can contact Ralph, Bruce, or Paul Simpson at *Classic Aviation*, Hangar 617, Bankstown airport, NSW. Tel: (02) 792 2288 or Fax (02) 708 5150.

Two recent imports have confirmed the **North American T-6** (in various guises) as the most numerically superior warbird currently operating in Australia. Both machines are **SNJ-5Bs**. The first, now registered as VH-OVO, was purchased by Mike Falls of Melbourne from California based Jim Mott, and is currently undergoing certification by *Aerotech* of Toowoomba. It will be painted into a U.S.M.C. colour scheme. The second aircraft, Bu Aer 90663, (ex N3633F) and now registered VH-USS has been acquired by Colin Rodgers of Sydney and will retain its overall yellow U.S. Navy colour scheme.

Canada

From Douglas Keall at **Cold Lake, Alberta,** comes news of a most exciting project. Nestling amongst the CF-18 Hornets of 410(OT) Squadron, CAFB Cold Lake is the de Havilland **Mosquito B.35 RS700** (ex CF-HMS) which is being restored to flying condition. The aircraft is almost complete, and a source of spruce has been located. Boeing Canada have offered assistance with the project, which is said to be long term, but many volunteers on the

Above: *From* Doug Keall *at Cold Lake, Alberta, comes this shot of dH Mosquito RS700 which is being rebuild by volunteers at the CAFB.* **Lower** *(Ian Roach) Ex Yugoslav Air Force P-47D Thunderbolt at* The Museum of Flying, *Santa Monica, California. The latter is not part of the auction!*

Colour Captions Opposite: Top (*Howard Pardue*). *Howard Pardue's immaculate R-3350 equipped racing Fury N666HP flying out of Breckenridge, Texas.* **Lower** (*Alan Gruening*) *The* Champlin Fighter Museum's *Fw190 running up in January 1990.*

Continued on Page 13

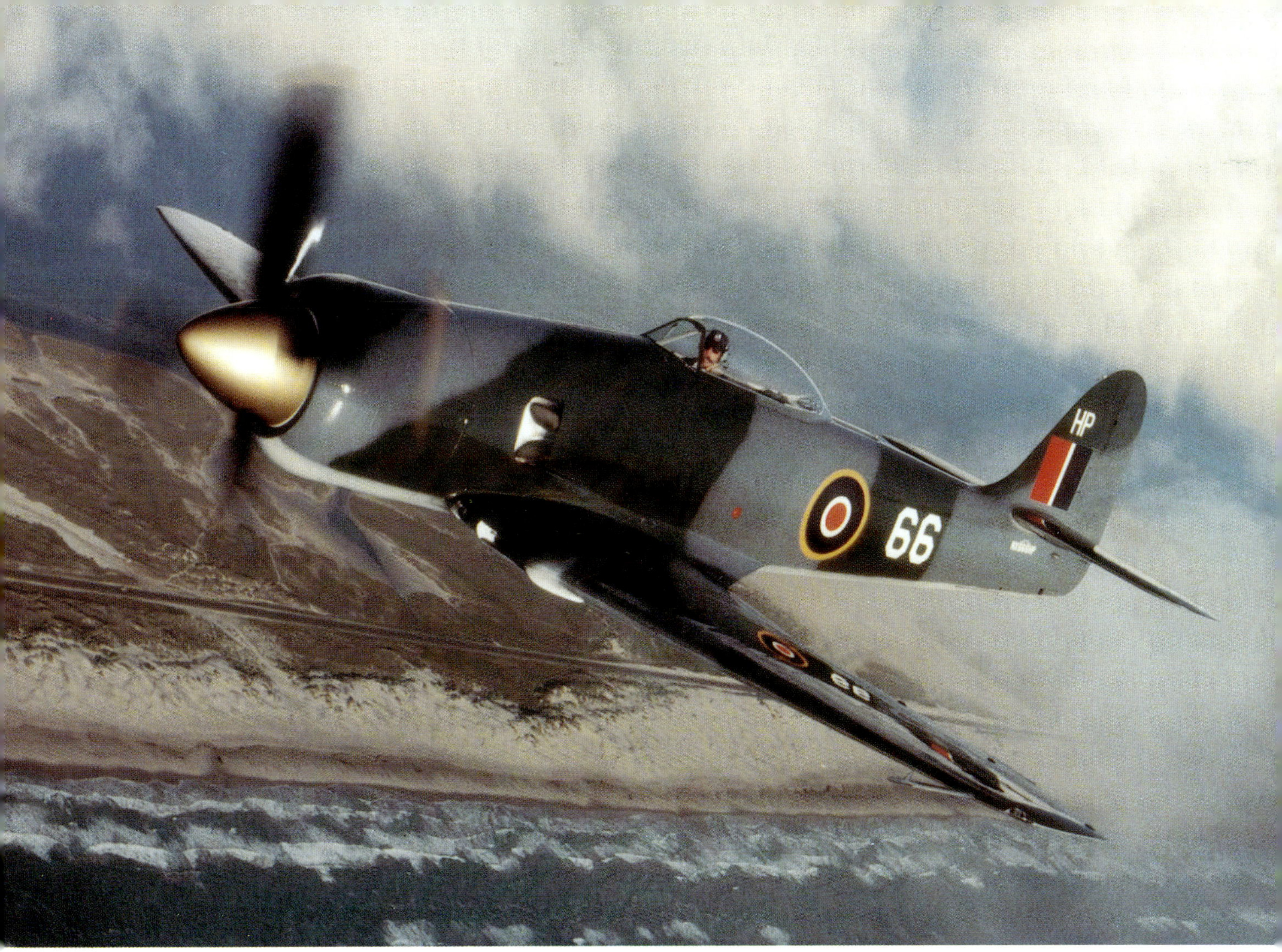
HP
66

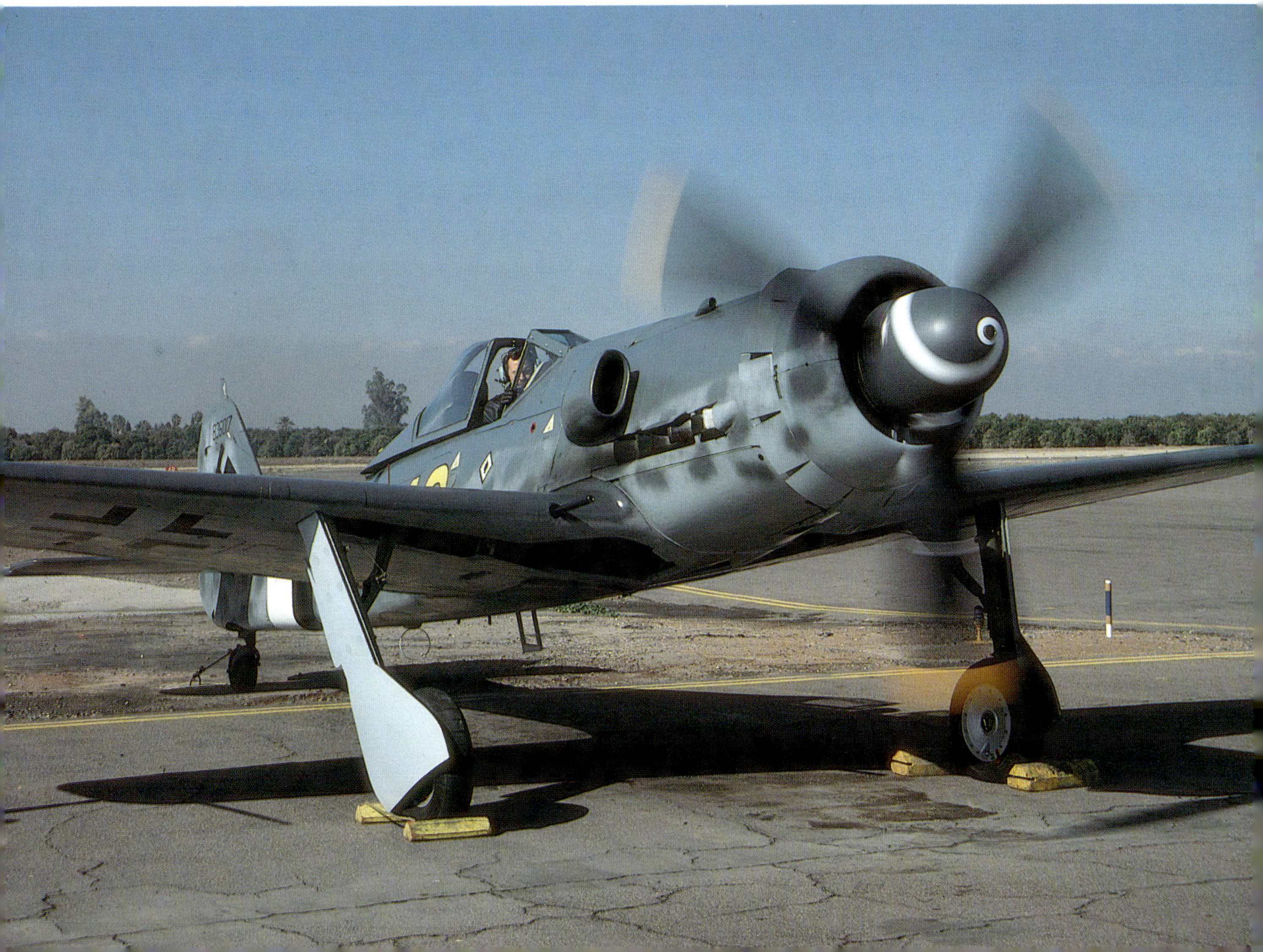

Airframe Assemblies Ltd
Vintage Aircraft Sheet Metal Specialists
WARBIRD COMPONENTS
Spitfire Specialists

Spitfire items currently being manufactured include;
Cowling sets (Merlin)
Cowling sets (Griffon)
Flaps
Ailerons
All fuselage frames
Rudders 5/9/14
Wing tips
Elevators
All fillet fairings
Aerial
Tailplanes
All longerons
Rad. fairings
Various hinges/brackets for rudder/elevator etc.
Complete rebuild of wings – in jig – including new spars
Spitfire Oil Tanks
Any other sheet metal requirements you may have for other vintage aircraft can be produced to drawings or pattern

Tel: 0983 404462

If you are contemplating an historic aircraft rebuild please contact;

**Unit 2,
East Yar Road,
Sandown,
IOW, PO36 9AY**

**Inspected and released.
We have produced parts for warbirds all over the world**

AVIATION JERSEY LIMITED

P.O Box 33, Rue des Pres,
St. Saviour, Jersey,
Channel Islands

PRODUCT SUPPORT FOR ROLLS-ROYCE MERLIN AERO ENGINES

Aviation Jersey Ltd., has the tools, equipment, publications and experienced staff to provide full technical support and overhaul facilities for Rolls-Royce Merlin engines. The Company holds a full range of spares. All spares are issued with an inspection certificate.

Contractors to British Ministry of Defence and other Governments.

TELEPHONE: 0534-25301
Fax: 0534 59449
TELEX: 4192161 AVIOJY G

Microscan Engineering Limited

▶ Specialists in C.N.C. and manual machining of Spitfire and all Warbird components

▶ We are experienced in the machining of awkward and hard to get components for <u>your</u> Warbird

▶ Let us help get your SPITFIRE, HURRICANE, MUSTANG etc. into the air more quickly ......

CONTACT
▶ **MARTIN EDWARDS**
for immediate personal attention on
0602 736588
or Fax
0602 461557

**ACTON CLOSE, ACTON TRADING ESTATE,
LONG EATON, NOTTINGHAM, NG10 1FZ, ENGLAND**

WARBIRDS WORLDWIDE SCENE

Continued from Page 10

base are offering assistance and the aircraft is undercover. We will be following the progress of this project as it develops.

Great Britain

The **Old Flying Machine Company** are adding another type to their inventory of warbirds with the acquisition of ex Egyptian Air Force **Yak 11** N11SN (featured on page 42 of WW Ten). Mark Hanna told *Warbirds Worldwide* that the aircraft should arrive in the UK in the near future. The high performance Yak was painstakingly restored in the United States and was previously owned by Neil Anderson.

A most interesting development on the UK scene is the transfer of a **Hawker Tempest II** ('HA457') from the RAF Museum to the *Fighter Collection* at Duxford for 'rebuild to static condition only' say the RAFM. Activity based on the Tempest in the UK has seen a recent upsurge and several aircraft have now been registered including *Autokraft's* HA604 (G-PEST) and HA586 (G-TEMT)

Work has been progressing on the cockpit of the *Fighter Collection's* Grumman Hellcat (see *Warbirds Worldwide 11* for an exclusive air to air shot of the aircraft) and the machine will soon be on its way to the UK.

Duxford have announced the dates of their *Classic Fighter Display* as Sunday, July 8th, and you will be able to visit the *Warbirds Worldwide* stand at this display.

United States of America

News was coming in as we went to press of the ditching of a North American **P-51D Mustang** off the Texas coast in January. The aircraft was en route from Miami, Florida to Corpus Christi, Texas, when it is believed the aircraft suffered fuel starvation and had to ditch in shallow water. The pilot escaped safely and it is understood there are plans to recover the aircraft.

An exciting event took place recently at the *Champlin Fighter Museum* at Mesa, Arizona, when the museum's **Fw190** Junkers Jumo engine was run up recently. Alan Gruening sent in a full report – see elsewhere in this edition.

The aviation event of the decade is due to take place at **Santa Monica,** California, on May 19th and 20th, when the **Museum of Flying** conduct their **auction.** See major feature on page six.

The Editor is always pleased to receive News items for inclusion in the SCENE section. Please address all correspondence to Warbirds Worldwide at the White Hart Chambers address on the Contents Page.

North American Mustang N4151D at the Champlin Fighter Museum *where it is currently hangared. John Sandberg (JRS ENTERPRISES - see article in the last issue) is in the cockpit performing engine runs* (Alan Gruening - **Top**) *Another shot of the Mosquito at CAFB Cold Lake, Alberta, shows the complete fuselage. (Doug Keall)* **below.**

Champlin Fighter Museum
Focke Wulf Fw190D Completed!

Alan Gruening looks at the pride of the *Champlin Fighter Museum*, and the recent completion of its restoration to running condition.

Ranked amongst the best fighter aircraft of World War II was the German Focke Wulf 190D series, some 700 of which were built by 1945 in the *Luftwaffe*'s effort to regain mastery of the air above its homeland. Only three 'long nose' 190s survived the scrap piles at the end of the war; two became property of the U.S. *National Air & Space Museum (NASM)* at Wright-Patterson AFB in Ohio, and at *NASM*'s Silver Hill, Maryland facility. The third of these rare machines belongs to the *Champlin Fighter Museum* based at Falcon Field, in Mesa, Arizona, and it has just been put through nearly a year of additional detailed restoration effort by the museum staff to bring it up to operating condition after almost 45 years.

Originally designed for the installation of a radial BMW engine, the Fw190 series went through several improvements to maintain an edge of superiority over many of its adversaries in the war and to become a fighter bomber for more effective close support work. Early in 1944 however a significantly different and technically superior version, the Fw190D came into production. In place of the radial engine the *Dora* series was equipped with an inverted V-12, liquid cooled Junkers Jumo engine which could generate 2240HP at full power! This radical change required major airframe modifications to accommodate, including the lengthening of the fuselage just ahead of the tail assembly, as well as extending the cowling to house the massive twelve cylinder engine out in front of the firewall. A unique feature was the cooling system, since the airframe designed for a radial engine had no provision for radiators, and it was desired to maintain the aerodynamic efficiency of the fighter without adding bulbous appendages in the airstream. The solution was to design a rounded radiator system that would fit around the nose cone of the engine under the ring cowling, retaining the appearance more typical of a fighter with a radial air cooled engine. The Focke Wulf design team, headed by Professor Kurt Tank produced a fighter which was an equal match for advanced Allied aircraft such as the North American P-51D Mustang. Maximum speed was rated at 426mph at 21,650 feet with a service ceiling of 32,810 feet. The *Dora* was armed with a 30mm cannon firing through the propeller shaft, two 20mm cannons in the wings, and could carry 1,100 pounds of bombs.

After the war, five of the long nose Focke-Wulfs were brought to the United States for flight evaluation. At the completion of this programme three of these aircraft were saved from the scrappers torch. Fw190 Nr. 836017 was donated to Georgia Technical University for engineering research. Subsequently it ended up in Santa Barbara, California, where it was neglected, abused and vandalized. However, in 1972 it was acquired by Doug Champlin who contracted with A. A. Williams in Guenzberg, Germany, to restore the rare fighter. Williams worked for four years, putting thousands of hours into the restoration at his facility. At that time, Professor Tank was still alive and he assisted Williams during the restoration process – a rare event indeed to have input available from the original designer of such a significant aircraft.

Returned to the United States in 1976, the Fw190 became the keystone and most valuable fighter in the *Champlin* collection. Essentially complete, the aircraft was put on display with the other famous fighters at the *Champlin Fighter Museum* when it was formed at Falcon Field in 1980. In addition to the 190D, Williams also converted a Spanish HA-1112 into the Museum's Me109E by the replacement of the Rolls-Royce

Top: Dave Goss, Chief mechanic at the Champlin Fighter Museum working on the Fw190s Jumo V-12 engine. Lower: Nearing completion of the massive restoration, test runs of the engine were made uncowled until proper adjustments were achieved.

Merlin with an original Daimler-Benz 601A powerplant. Lead by the museum's Chief mechanic, Dave Goss, the Champlin crew performed a detailed final restoration on that aircraft in 1988 and successfully ran the DB-601 engine, though the aircraft was considered too valuable to risk actually flying it.

Following this success however, the decision was made to repeat the achievement with the Fw190D-12 in 1989. Early in that year the aircraft was taken from the display hangar and moved into the restoration hangar at the museum's facility. For nearly a full year Dave Goss and Charlie Hyer disassembled, inspected, reconditioned and repaired all systems of the aircraft to bring it into operational condition. The big Jumo engine was opened for inspection of crankshaft bearings, lubrication system, fuel, ignition, and cooling systems – everything necessary to bring it back to life again. Art

Continued on Page 46

Ill fated Sea Fury NX62143 which was destroyed in a hangar fire at Shafter in July 1988 (Alan Gruening)

Sea Fury

Graham Trant examines the latest happenings on the Sea Fury scene and plans for the future that will ensure the type's survival well into the next century.

The year 1989 was not a good one for the Hawker Sea Fury. With the loss of the *Fleet Air Arm Historic Flight's* first production FB11 TF956 off the coast of Scotland on the 10th June, thankfully without the loss of the pilot Lt. Cdr. John Beattie RN, the scene was set. Although initial plans called for the recovery of the remains of the airframe, Royal Navy divers, upon locating the wreck reported the largest identifiable part was the mangled Centaurus engine. A few small items were brought to the surface, but there the activities ended. A malfunction of an undercarriage 'up lock' is thought to have been responsible for preventing the leg from lowering and staying firmly in its well.

The abandonment of the aircraft was, at the time, a controversial issue and many opinions were voiced. Should the pilot have attempted a landing, perhaps on a foam carpet at an RAF Master Diversion airfield? However, for the pilot, instructions were quite clear and when the regulations are as plain the pilot has no option. Apart from the 'distinction' of leaving a warbird he is no doubt now a member of both the *Caterpillar* and *Goldfish* clubs! Although some of the racing Furies in the USA and Canada have made successful one wheel landings, one at least on foam, when flying one of Her Majesty's aircraft the rules must be obeyed. In my opinion Lt Cdr Beattie took the right action: after all a structure of steel and aluminium pales into insignificance compared to a human life. TF956 now rests in many thousands of pieces at sea off the Scottish coast, or does it? While the aircraft did indeed go into the sea on that fateful June day, parts of the original T956 still exist here in the UK and will, in due course, fly again.

In 1974 TF956, during landing following a display at its home base at RNAS Yeovilton, suffered a partial undercarriage collapse, damaging the starboard outer wing and centre section, as well as the engine and propeller. The engineers at Yeovilton repaired the aircraft utilising parts from the *FAA Museum's* FB11 serial WJ231, as well as installing a spare wing and centre section. It was reported later that this rebuild incorporated the centre section from the ex German Sea Fury T.20 D-CIBO (ex D-FIBO and VX309), although these did not in fact arrive in the UK (with the *FAAHF's* T.20 WG655 (D-CACU)) until 1976, sometime after the single seat TF956 flew again. The centre section from TF956 and its damaged wing were stored by the *FAAHF* at Yeovilton until 1989 when they were purchased by Essex based engineer Craig Charleston.

Craig is well known for his high standards of workmanship, in particular on the Spitfire FR XIVc NH749/NX749DP rebuilt for the late Keith Wickenden, and now flying with David Price at the Santa Monica *Museum of Flying*. He has been collecting Sea Fury parts for some time and is now able to identify two very clear, but long term restoration projects.

The well known Ormond Haydon Baillie Sea Fury WH589, alias CF-OHB and G-AGHB, is not, as has been published many times, in the hands of Lloyd Hamilton at Santa Rosa in California. Whilst it is true that some parts (the rear fuselage in fact) were incorporated into the rebuild of *Furias* (registered as N4434P), that basic airframe is that of an ex *Kon-Marine* Sea Fury FB50, built as VX715 for the Royal Navy and later flown from the Dutch carrier *Karl Doorman* coded 6-14 and later 10-14. The rear fuselage of the Dutch Fury was damaged, and it was easier to substitute the section from the then recently damaged G-AGHB which was available following its accident at Munster in West Germany. So *Furias* is in fact the rebuild of this Dutch Fury, with parts from other aircraft including this tail cone from the well known Haydon Baillie airframe. Craig Charleston intends to rebuild

G-AGHB using its original cockpit section, fin, and tailplane, but incorporating the centre section from TF956, the rear fuselage from VX715 (the ex Dutch aircraft) together with brand new outer wings, rudder, elevators etc. A powerplant has not yet been earmarked as the project is long term, but both Centaurus engines and Sea Fury propellers are available here in the UK. Craig advises that the Civil Aviation Authority have been very helpful and he hopes that the registration G-AGHB can be restored for this aircraft in due course. (*Editor's note:* The United States civil registry clearly states that the identity of Lloyd Hamilton's Sea Fury is WH589, and has for several years. The registration G-AGHB was, according to the Civil Aviation Authority, formerly allocated to an Avro Anson, then taken up by WH589 in May 1984. It was finally withdrawn in September 1981, still in the name of O. Haydon-Baillie). All parts for this rebuild are in good condition, even those from the damaged original G-AGHB. This had passed to the ownership of Spencer Flack following the untimely death of Haydon-Baillie in the Cavalier Mustang I-BILL in the summer of 1977. Spencer Flack obtained the aircraft after it had stood dormant for sometime and following maintenance at Munster embarked on a test flight in June 1979 with Mike Searle at the controls. Fuel flow problems forced him to make an emergency landing during which repairable damage was sustained. Spencer had WJ244 on rebuild at Elstree, later to emerge in the now famous Flack scarlet colours as G-FURY. This aircraft appeared at a number of shows in the late 1970's and early 1980's before crashing at Waddington in August 1981. Whilst the Fury was bodily damaged and burned following the post crash fire from which Spencer escaped, many of the ancilliary parts were left in good condition. There the story should have ended with the wreckage passing into the hands of the scrapman, only to be rescued by a spare parts collector. The Fury remains were stored and later passed into the hands of Craig Charleston along with those of G-AGHB, and other miscellaneous Sea Fury spares. For the moment however Craig is busy rebuilding Spitfires and the Furies are future projects.

The Royal Air Force's *Aircraft and Armament Experimental Establishment (A&AEE)* at Boscombe Down have now completed the repairs to their ex German Sea Fury T.20 VZ345 following its unfortunate accident in spring 1985 when it flipped over onto its back. The main area of damage was to the fin, twin canopies and turtle decking as well as the engine and propeller. For sometime the future of this low houred Sea Fury was in doubt. However, in 1988 work commenced on its rebuild, with the assistance of Craig Charleston. The Royal Navy were also very helpful and Sea Fury FB11 VR930 (ex RAF Colerne museum and Wroughton) was loaned to Boscombe for spares use and as a pattern airframe. The engineers at Boscombe Down have performed a fine restoration and whilst using some parts from VR930 and the rudder from G-FURY, the bulk of the repairs have been effected on the original airframe. The aircraft was painted in the autumn of 1989 and has recently been engine running prior to an imminent first flight. VZ345 was used by A&AEE as a high speed photographic aircraft for filming air drops from Royal Air Force C-130s and on other test work, and not as the popular press would have you believe as the Station Commander's run-about! In addition to its photographic missions the aircraft was also used to give tail dragger experience to *Battle of Britain Memorial Flight* pilots in conjunction with the Boscombe based Harvards.

Robert Lamplough's Sea Fury TG114/N232J is (as we go to press) in Canada awaiting clearer weather before continuing its journey to North Weald. Robs raced this aircraft at Reno in 1988, his fine feature *Reno Rookie* appearing in *Warbirds Worldwide* Number Seven.

At Biggin Hill, *Warbirds of Great Britain* have WJ288/G-SALY which is in good condition having been partially restored by its previous owners.

Part Two of this article will feature a listing of all known existing Furies together with the latest news on this most exciting and increasingly popular warbird.

WW Graham Trant

WARBIRDS WORLDWIDE

Frank Sanders in Sea Fury N924G rounding the pylons at Reno last year. The Sea Fury is becoming more and more popular with the remaining Iraqi Fury airframes being snapped up. (Thierry Thomassin)

Flying Sugarland Express

For those unfamiliar with the Invader it is a twin engined attack bomber built by Douglas. Without going too deeply into the aircraft's history, which is beyond the scope of this article, the Invader resulted from the U.S. Army Air Corps requirement of 1940 for a multi-role light bomber with a fast low-level capability with an alternative deployment from medium altitude for precision bombing attack, and a heavy defensive armament. To meet the required specification Douglas proposed a developed and enlarged version of the A-20, powered by P&W R2800 Twin Wasps. Some three prototypes were ordered in June 1941, and the first XA-26 made its maiden flight on 10 July 1942.

After extensive service trials it quickly became evident that Douglas had built an aircraft more than capable of handling its intended role: indeed the A-26, exceeded every performance specification.

The potential of the A-26, coupled with its superb performance ensured that it continued in service with the USAF for many years after World War II. The Invader served as an early NATO reinforcement in Europe, in Korea with great distinction, and saw combat in Vietnam in 1962. Many foreign air arms utilised the type and several were later modified for utilisation as executive transport aircraft. Out of the 2500 aircraft built a surprising number survive today (see the *Warbirds Worldwide Directory* for a full listing – Ed). To sum up the Invader in a sentence, it is a magnificent aircraft with superb handling characteristics, an impressive payload capability and high speed performance.

The maximum military weight of the B model is 35 000 lb, net weight 22 683 lb,

Anders K. Saether, the driving force behind the Scandinavian Historic Flight, gives an insight into flying this powerful Douglas twin engined bomber.

which gives a useful payload of 12 638 lb. Wing span is 70 feet, length 51 feet 3 inches and the height 18 feet 3 inches.

I first saw an Invader fly in 1958 when *Wilderoe Flyveskap* in Oslo used the aircraft for photographic recce missions over Svalbard. This aircraft was an A-26B model – surplussed from a USAAF depot in Europe. I had never seen an aircraft with such speed and grace before. The impression stayed with me, and my desire to own an Invader became stronger and stronger.

It was not until November 1987 that the chance to fulfill my dream arose. I purchased an A-26B, at the time registered N8392H, serial 44-34602. This particular aircraft was built as a trainer and never served in combat. It had some 4 500 hours total time and had never been damaged. In 1961 it was converted to executive configuration as a Monarch Invader, and was owned by a company in New Mexico for all of its civilian career until I purchased it.

West Star Aviation of Grand Junction, Colorado, performed a total restoration and it was painted in a 729th Bomb Squadron colour scheme of the Korea era, and named *Sugarland Express*. Eight 0.50 in guns were installed in the nose, and the aircraft was re-registered N167B.

I now found myself in the position of being the proud owner of an Invader without even having a multi-engined rating. My flying background of some 2 000 hours was strictly on single engined aircraft, with about half this time on high performance warbirds.

My instructor, Tom Dollahite and I flew the Invader approximately ten hours in preparation for my twin rating and instrument type rating. Needless to say it was a very nervous candidate who presented himself for the FAA check ride. After a one hour flight and lots of perspiration I got it!

The 4 700 nm trip home to Norway (in May 1988) was not uneventful. Our route was Grand Junction Colorado, Oklahoma City, Indianapolis, Buffalo, Sept-Iles (Canada), Goose Bay, Kuujjuaq, Frobisher Bay, Sondre Stromfjord (Greenland), Reykjavik Island, Bergen and

finally Oslo. We had temperatures down to minus 40 degrees, strong winds, and IFR conditions. In addition to this our heater gave out passing over Montreal northbound and we were never able to make it work. The *Hudson Bay Company* at Frobisher Bay had some good business in arctic clothing from us, and we looked like stuffed teddy-bears! We learnt a lot from our safety pilot for the trip, Steve Kehmeier.

After having flown the Invader in Europe for two seasons I now have a much better idea of what to look out for in the operation of the aircraft. In my opinion, getting the aircraft stopped *on* the runway after touch down represents the biggest challenge in flying the A-26.

Cockpit entry is through the tunnel in the fuselage. Emergency exit is over the top of the wing. The view from the cockpit is excellent. The shoulder wing makes it possible to look into the turn as well as down.

Our take off procedure is usually

Lars Ness brings the SHF Invader around in a tight turn at the Warbirds Worldwide meeting at Duxford in June 1989 (**Richard Paver**)

different from that adopted by the military. Normal take offs are flapless. Runway length and terrain permitting, we keep the aircraft on the ground until we have reached 130 knots. At sea-level this will take 1 200 metres with a medium load on a standard day. The gear comes up instantly as soon as positively airborne. We hold the nose down and accelerate to 180 knots and then control the climb to reach 500 feet at 200 knots. At that point we make the first power reduction from 52 inches to climb power which is 42 inches and 2 400 rpm. If one engine quits before we have reached 130 knots we abort the take off. If necessary, we will over-ride the ground, look for the gear up position and belly the aircraft. On short runways (1 100 metres hard surface minimum) we use 20 degrees of flap and rotate at 100 knots. We allow essential crew only on board.

Handling characteristics are excellent. Elevator and aileron controls are light at normal speeds. The aircraft has a gentle stall with a distinct buffet, which occurs at 96 knots at 26 000 lb clean. At 2G, the same weight, the stall occurs at 139 knots with considerably less buffet and sometimes a wing drop. At 3G the stall occurs

at 170 knots. At a speed above 300 knots the elevator is still light, but the ailerons stiffen up. The airframe is stressed for 4.7G and −2.3G at a maximum of 29 000 lb. As an instrument platform the aircraft is extremely stable. In calm air and properly trimmed *Sugarland Express* will fly hands off for several minutes.

At an entry speed of 340 to 350 knots and a gentle initial pull of +4G the aircraft will still have 150 knots at the top of a half cuban. You will gain approximately 5 000 feet in the manoeuvre. At times our Mustang has to use high power (50″+MP) to stay in formation. Roll rate is 30 degrees per second. In a climbing roll with an entry speed of 300 knots you will gain approximately 2 000 feet of altitude and exit at 150 knots. Maximum manoeuvering speed is 240 knots.

The Invader is extremely fast. *Sugarland Express* does 300 knots IAS at sea level, max continuous. Cruise performance at 20 000 feet and 55 per cent power is 340 knots true. She will then burn 150 U.S. gallons/hour giving 5.3 hours total endurance out of 800 U.S. gallons fuel capacity.

Needless to say, we use a detailed check list at all times. We also have detailed procedures dividing the work load between flying and non-flying pilot. Displays are done with minimum crew on board and preferably with not much fuel in the outboard main tanks. When carrying passengers (8 maximum) the manoeuvering of the aircraft is limited to airline standards.

Modern conveniences like steerable nose-wheel, anti skid brakes, reverse thrust propellers and spoilers do not exist. Combined with the touch down speed of 100 knots and a weight of approximately 2 700 lb, there is a lot of momentum to kill and keep under control. It feels like sitting in the locomotive of a freight train with 50 fully loaded cars pushing.

The most efficient way to stop is to touch down three to five knots above stall, apply full up elevator and use the 70 foot wing as a speed brake for as long as you can keep the nose up. When the nose wheel touches you select flap up and apply brakes as soon as the aircraft sits heavily on the runway. Using this technique the aircraft will stop in approximately 800 metres on a dry runway with no wind.

Strong cross-wind landings require a long (at least 1 600 metres) dry runway. The flight manual does not give a maximum demonstrated cross wind component. We have, however, decided to use 15 knots as our limit. This may seem rather low, but believe me, you will have your hands full in keeping the aircraft on the runway. The large area of the tailfin (shown to advantage in the colour centre spread of this edition – Ed) and the long fuselage makes the A–26 weather vane like a tail dragger. Sometimes use of the up wind engine is required in addition to full rudder and down wind brake. It is extremely easy to lock a wheel and blow a tyre, in which case you are on for the ride.

You can apply approach flaps (20 degrees) at 180 knots. Full flap and gear limitations are 140 knots and final at 120 knots. Full flap is not selected before landing is assured. Our VFR approaches are high as VMC is 130 knots; it is nice to know you can glide to safety. Single engine go-around cannot be performed under 500 feet so at this point we consider ourselves committed. Performing ILS approaches to low minimas are not pleasant as you fly low and below VMC for a long period before touch down. As a result our IFR minimum is 800 feet.

From a maintenance point of view *Sugarland Express* has performed excellently. The most serious snag we have had in some 200 hours of flying are some loose exhaust stacks, an inherent problem of the A-26. *Sugarland Express* is all that we had hoped for. A popular aircraft at displays and a fantastic mother-ship for the *Scandinavian Historic Flight*.

WW Anders, K. Saether.

Missing an Edition?

BACK NUMBERS

Direct from the Publishers at Post Free prices. Number One – Sold Out.
Number Two and Mustangs Worldwide – Low stocks. Sent airmail overseas. UK £3.95 (Number Nine £4.50), Europe £4.50 (Nine £5.00), U.S.A. $7.95, Australia AUS$9.00, New Zealand NZ$9.00, Canada CAN$9.00. Sent Airmail overseas.

WARBIRDS WORLDWIDE, FIVE WHITE HART CHAMBERS, 16 WHITE HART STREET, MANSFIELD, NOTTS NG18 1DG, ENGLAND.

The Scandinavian Historic Flight

VIKING SQUADRON

The Scandinavian Historic Flight

The *Scandinavian Historic Flight* (SHF) was formed with the aim of restoring, maintaining and displaying vintage military aircraft. The group believes in the importance of preserving the heritage of military aviation. This can only be achieved by making it possible for people of this and future generations to view and experience the aircraft in their true element.

To achieve this determination and a lot of hard work is required. Aircraft become more and more difficult to find and increasingly expensive to maintain and fly. In order to achieve these goals the *SHF* is dependant upon the support and co-operation of many people. Firstly the people who pay entrance fees at airshows to see warbirds. Secondly their members who pay NKr.100 (£10) and help *SHF* at every possible opportunity when practical assistance is required; and last but not least their active members who share the responsibility of the running of *SHF Oslo Wing*.

SHF have a clear policy on safety. Their aim of flying the aircraft is never undertaken at the expense of safety. This philosophy begins on the ground. No compromise is made in repairing any mechanical malfunction if it affects flight safety. The aircraft are grounded if they cannot be ready in time to undertake a display. Excellent routine maintenance prevents almost all serious snags. However, with the operation of World War II aircraft (even though they are of excellent design and manufacture) breakdowns do sometimes occur. Fortunately this almost exclusively happens to components non-critical for keeping the aircraft in the air.

Of equal importance is the vetting of

Paul Coggan looks at the *Viking Squadron* of the Scandinavian Historic Flight in this Warbirds Worldwide special feature.

Anders Saether, the driving force behind the SHF, *in the cockpit of N167F* (Mette Lium).

aircrew carried out by the *SHF*. Pilot experience, currency and discipline is paramount. the aim is to execute every display on time, in a disciplined manner and with flight safety as a paramount consideration. "SHF . . ." says Anders Saether . . . ."is not a training ground for pilots who want to show off their skills at the weekend"

Although piloting is a never ending learning process, the SHF pilots have all proven themselves elsewhere. The combination of skill and maturity is an absolute requirement. Pilot display limitations are individual and set with good margins. All displays are briefed and executed with military precision and always as planned and approved prior to the display.

During the 1989 display season the *Scandinavian Historic Flight* undertook 48 displays. Additionally the Flight's North American P-51D Mustang took part in the filming of the David Puttnam movie *Memphis Belle*. The *SHF* have a close working relationship with the UK based Old Flying Machine Company, and when enquiries for aircraft other than the types they operate the OFMC are contacted.

SHF are a professional organisation with professional engineers, crews and operations staff. They have an excellent fleet of reliable and well maintained aircraft, all equipped to handle the most difficult weather conditions. Plans are afoot to acquire other aircraft for the *SHF* and we look forward to seeing them display in Scandinavia and Europe for many years to come.

The Aircraft

Noorduyn (AT-16) Harvard IIB LN-TEX/KF568

Built by Noorduyn in Canada KF568 served initially with No. 6FTS at Little Rissington, later moving on to No.1 squadron RAF which was then based at RAF Tangmere. In late September 1949 KF568 was transferred to the Belgian Air Force and serialled H-58. The aircraft was declared surplus by the Belgian Air Force and taken on by a target-towing company and registered OO-AAR. The aircraft was later based in West Germany and registered D-FIBU. The Portuguese Air Force were the next recipients of the Harvard, and they dubbed it FAP 1794. The aircraft is known to have seen combat in Angola, carrying gun pods and rockets. The aircraft took up the registration LN-TEX in August 1979 and has provided excellent service since.

North American T-28 Trojan Bu. 140547/ N2800Q

This ex U.S. Navy T-28C is believed to be one of the earliest 'C' model Trojan warbirds. The *Scandinavian Historic Flight* acquired the aircraft through Denny Sherman in 1986. The aircraft had been totally restored with no expense spared, and had been fitted with new engine, new propeller, new wing mounts, all new landing gears and high stress components and a new canopy. The aircraft was ferried to Colorado and then out to Norway in the spring of 1986

North American P-51D Mustang 44-73877/N167F

Built at the Inglewood facility of North American Aviation, P-51D serial 44-73877 was delivered to the USAAF in mid 1944. It spent the majority of its service life with Stateside training units before entry into service with the Royal Canadian Air Force on 23rd January 1951. Dubbed '9279' with the RCAF the aircraft served with 403 *City of Calgary* Squadron, based at Calgary, Alberta. After an accident free career with the RCAF it was struck off strength on 29th April 1959. After disposal to *Defuria & Ritts* by the Crown Assets Disposal Corporation the aircraft was put into storage for a brief period before being sold to *Aero Enterprises* at Elkhart, Indiana.

Neil McClain of Strathmore, Alberta, purchased the aircraft from *Aero Enterprises* just a few months later, registering it as CF-PCZ. On 29th April 1968 the Mustang was sold to Paul D. Finefrock of Hobart, Oklahoma, who registered it N167F. By the end of October 1970 Finefrock had flown the Mustang to Brownwood, Texas, where it was based with *Gardner Flyers*.

The SHF *Harvard IIB LN-TEX taken over Oslo Fjord.*

In 1974 N167F was flown to *Vintage Aircraft Limited* of Fort Collins, Colorado, ostensibly for restoration. After some time Finefrock decided to sell the aircraft. Darrell Skurich, the owner of *Vintage Aircraft,* telephoned Anders Saether who had been looking for a Mustang for some time. The undamaged and almost stock Mustang was exactly what Anders was looking for; it was in excellent shape, had never had a major accident either in military or civilian use and was free from corrosion. It was however in need of a total restoration, the crew at *Vintage Aircraft* started to dismantle the aircraft under Skurich's supervision. N167F was broken down

The SHF *T-28C Trojan at Oslo Fornebu airport.*

Centrespread: SHF's *mother-ship, the A-26B-61DL Invader 44-34602, captured from the ramp of a C-130 Hercules during the filming of a Royal Norwegian Air Force PR video in 1989. Lars Ness and Rolf Meum are at the controls.*

The Scandinavian Historic Flight

VIKING SQUADRON - Ovre Ullern Terrasse 27
0380 OSLO   NORWAY
Tel: 472 502365   Fax: 472 521489

Scandinavian Historic Flight - VIKING SQUADRON

into the smallest possible components. Every one was cleaned, inspected and where necessary replaced. Once empty the fuselage was completely stripped of paint. Some skins on the tail cone were replaced and some of the wing and cowling fairings were replaced. Light weight, foam-filled tanks were installed and a baggage door fitted on the port side. The seats were fitted with leather upholstery.

In line with *SHF's* flight safety policy the aircraft was designed to be externally authentic, with a modern cockpit fit including a new instrument panel. Full IFR capability was incorporated into the aircraft and a three axis autopilot was fitted plus an extensive modern avionics fit.

Jack Hovey of *Hovey Machine Products* was selected to rebuild the V1650-7 Merlin which was fitted with transport heads and banks, balanced crankshaft and other modifications.

After the flying the aircraft was flown to Arden Fisher in San Antonio, Texas, for painting. By July 1985 the job was complete and the paint scheme – that of Captain Bud Anderson's *Old Crow* (Chuck Yeager's wingman in the 357th FG) was applied, with the code *B6-S*.

Spring 1986 saw Anders Saether checking out on the Mustang in the U.S.A. prior to ferrying the aircraft to Norway. After spending some time in the U.K. on the display circuit the same summer, N167F was flown to Norway. It took part in the making of the film *Memphis Belle* in 1989 and has recently been repainted in its 357th FG *Old Crow* colour scheme.

Anders Saether in P-51D Old crow *acts as wingman to* the Old Flying Machine Company *Spitfire Mk IX MH434 with Mark Hanna at the controls.*

Douglas A-26B Invader
44-34602/N167B

Built by Douglas at Long Beach, California in 1944, the *Scandinavian Historic Flight's* A-26B served with several 'Stateside based training units before being declared surplus in the late 1950s. Following retirement from the military (in 1961) and a period of storage the aircraft was modified to *Monarch 26* standards with an executive style fit, long nose, airstair and custom avionics. It was later purchased by New Mexico based *Stahman Farms* and registered N8392H.

Anders began his search for an Invader in 1987, enlisting the help of Russ Williams and Bob Batterman. Bob has a wide experience of the Invader in USAF service. After being alerted to the availability of '602 and having inspected the airframe, Anders purchased it in November 1987, at which time it had flown 4300 hours total time and had never been damaged. *Stahman Farms* had looked after the aircraft, it having been hangared when not in use. Anders decided that *West Star Aviation* of Grand Junction, Colorado, should restore the aircraft. Under the supervision of Russ Williams, *West Star* installed an eight gun nose, a military style clamshell canopy and all new cockpit glazing, overhauled propellers and carbs, fitted dual brakes in the cockpit, installed King Gold Crown avionics and performed a complete D check inspection on the airframe which was then repainted in 729th Bomb Squadron, 452nd Bomb Group markings of the Korean war era. Before the flight to Norway the aircraft was re-registered N167B. It has since been christened *Sugarland Express*.

The Scandinavian Historic Flight People

SHF *Pilots from left to right: Rolf Meum, Jan Skjoldhammer, Lars Ness, reserve pilot Arve Wilmann and Anders Saether.*

The *Scandinavian Historic Flight* use four main pilots. All are highly qualified and experienced in many fields. Anders Saether is the force behind *SHF*. Anders is a self employed businessman and a Graduate (PMD) of Harvard University. He gained his PPL in 1962 and has since qualified with a commercial and multi-engined instrument licences. Anders has seaplane, ski and amphibious ratings as well as an aerobatic display rating and an unlimited low-level waiver. His most recent qualification has been the Douglas Invader instrument type rating. He has flown the Atlantic in both the Mustang and Invader and has been performing displays since 1980. He has well over 2000 hours total flying time, with more than half of this in high performance warbirds including the T-6, T-28, P-51 and A-26.

Rolf Meum is currently a DC-9 instructor and pilot with *Scandinavian Airlines System (SAS)*. He obtained his PPL in 1976 and went on to a distinguished military flying career, graduating as best in his class at primary flying training. He earned his American and Norwegian wings after training on T-37s and T-38s in the U.S.A. and graduated number one in his class with an exceptional rating. From 1979-1987 he flew F5s in Norway with the RNoAF and instructed at the *Euro-NATO Joint Jet Pilot Training* school in Texas. His aerobatic experience includes displays with Pitts Specials, T-6, T-28, and the P-51. Rolf has some 4000 flying hours in 59 different types and his experience in warbirds includes flying the F4U Corsair. 1700 hours are on jet fighters. He has an unlimited low-level waiver, and is a military instructor and Flight examiner. Both Anders and Rolf were in the UK in 1989 with the *SHF* Mustang, flying for the filming of *Memphis Belle*.

Lars Ness is a commercial pilot with the Dutch airline *Transavia* with over 2000 hours total time, half of which was as an agricultural pilot. He has a wide experience as an aerobatic instructor and has flown aerobatic competitions and displays. Lars has flown the T-6, P-51 and the Invader.

Jan Skjoldhammer gained his PPL at the age of 20 and has over 2000 hours total time, 1500 hours on twins. He is a Commuter airline pilot flying the F-50 for *SAS Commuter*. He gained his Commercial, twin and instrument licences in the United States and has an Economics degree from the Norwegian School of Management.

The operations section of *SHF* also perform a vital role, keeping track of the airshow scene throughout Europe each season, and booking the *SHF* aircraft into as many events as is possible. David Hammond, Mette Lium and Morten Myhr perform this task.

Also performing a vital task are the engineers of the SHF. They keep all the aircraft meticulously maintained throughout the year and fly with the Flight to shows to perform routine maintenance tasks and troubleshoot where necessary in the field. SHF engineers are Henrik Wideroe, Frode Johannessen, Tor Norstegaard and Lars Hansen.

The *Scandinavian Historic Flight* are a unique team of people. Anders Saether makes no bones about the fact that each member of the flight is a vital part of the professional operation, of which they are all justifiably proud.

Aircraft of the *Scandinavian Historic Flight Viking Squadron* are available for airshows, media and filmwork throughout Europe. If you are interested please contact them on 472 502365 or Fax 472 521489 (Norway)

Back cover captions: Top: *North American T-28 N2800Q over the Norwegian mountains with Rolf Meum at the controls.* **Lower:** *Anders Saether at the helm of Harvard IIB LN-TEX.*

SHF *Operations Staff: (Left to right) David Hammond, Mette Lium, and Morten Myhr.*

SHF *Engineers: (Left to right) Henrik Wideroe, Frode Johannessen, Tor Norstgaard and Lars Hansen.*

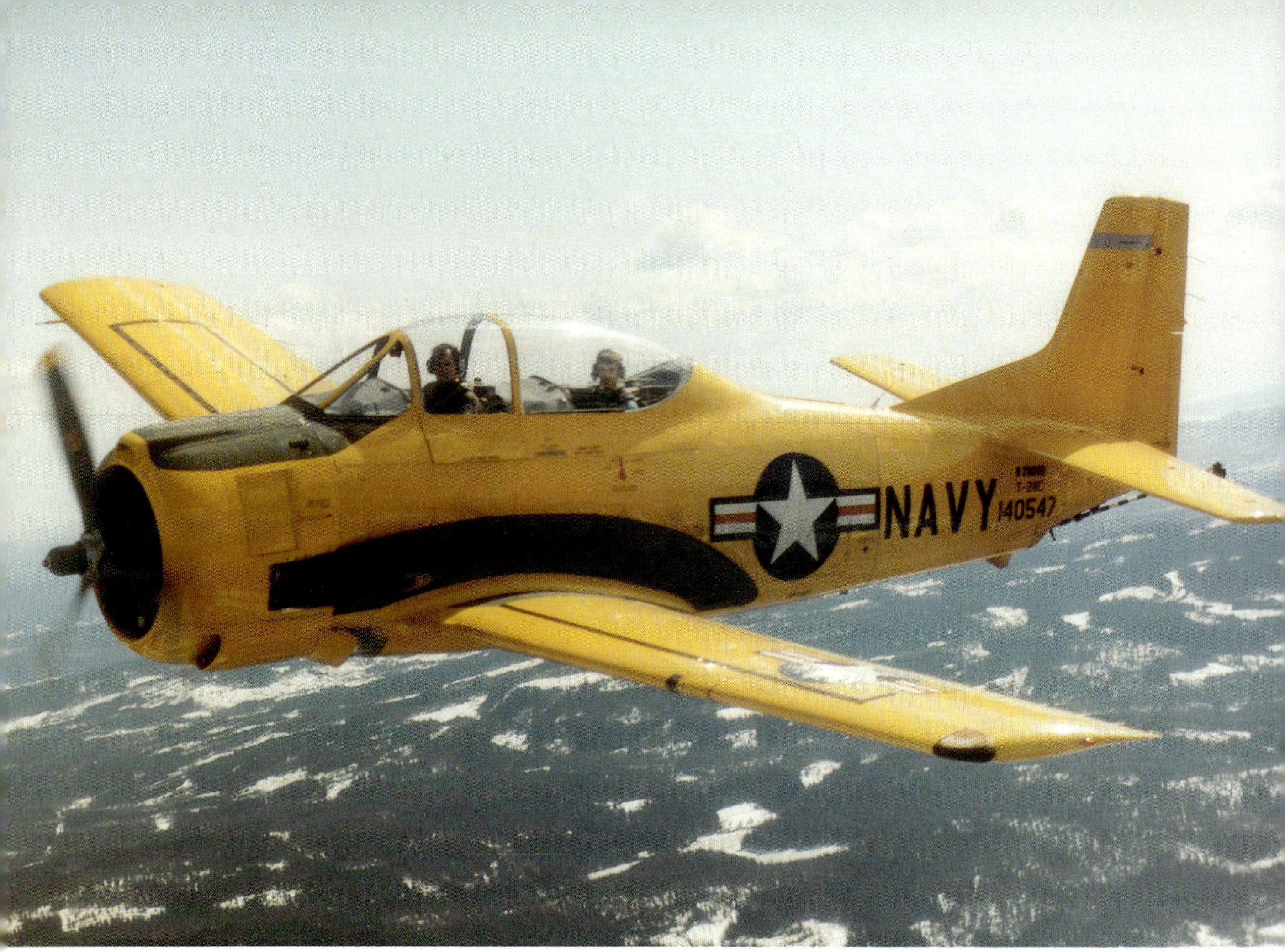
NAVY
140547

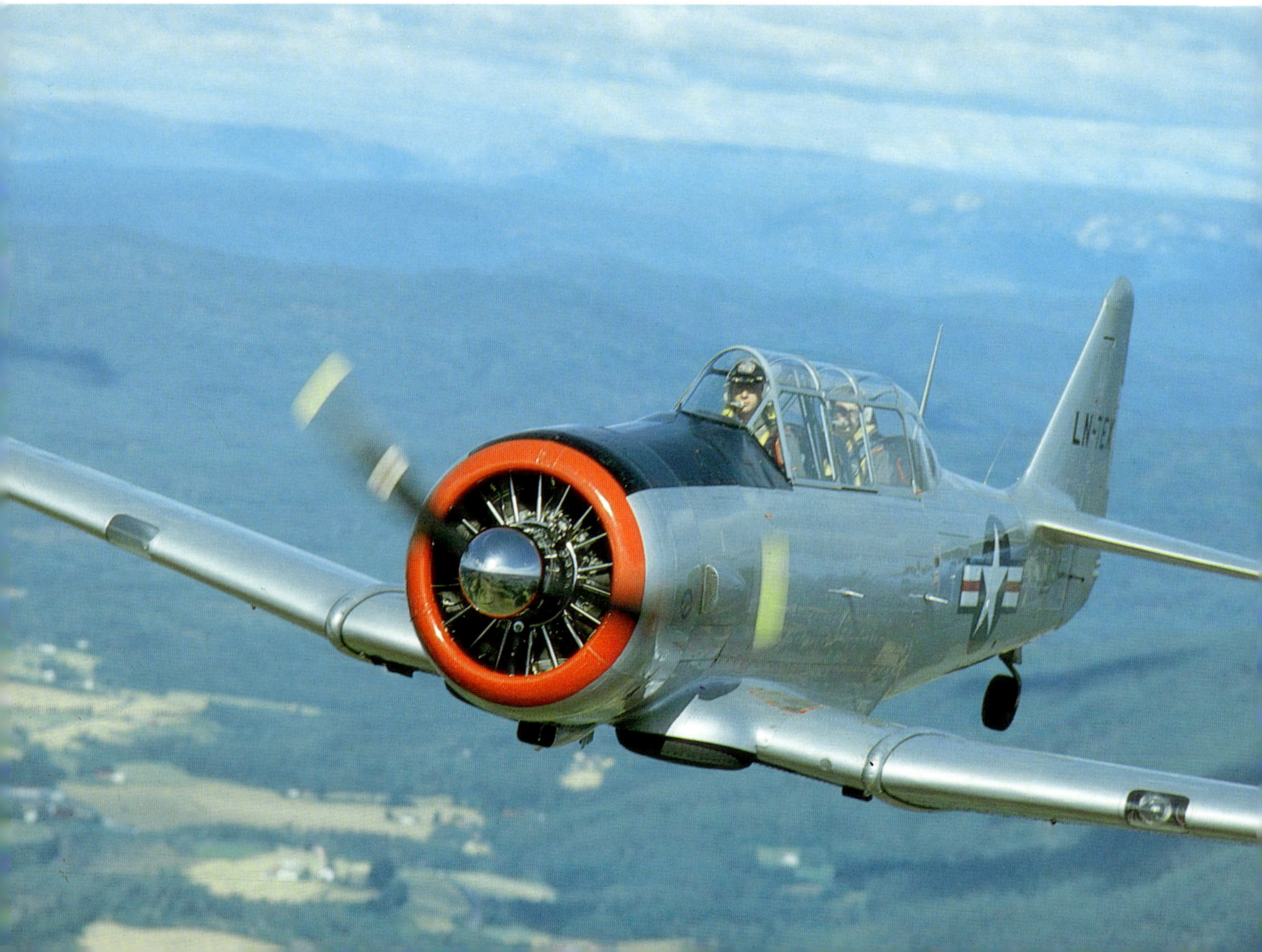

The Terrific Trojan

Walking up to the T-28, the first thing you notice is the size of the machine. This big radial engined aircraft, sitting on its tricycle gear, is tall enough to let you walk upright under the outer part of the wings. The fin is over 4 metres high. It really is quite big, for a basic trainer.

Our T-28 is a weapons modified C model, which means it's got the big engine, tailhook, and 6 hardpoints under the wing. The airframe has been beefed up to withstand carrier operations and weapons delivery

To enter the Trojan you have to lower the flaps. You do this by pulling down on a lever mounted on the left side of the fuselage just above the trailing edge of the flaps. With the flaps down you climb onto the wingroot, using footsteps in the flap and handholds in the fuselage and wing. Once on the wing you unlock the canopy and push it open. A word of caution; be careful when using the canopy lock/unlock lever as this also controls the emergency opening system. Upon activating this, the canopy goes to the full open position in approx. 1/2 second. *This can cause injury* to anyone in the proximity of the canopy! With the canopy open you enter either cockpit by using a footstep in the fuselage side.

Rolf Meum has flown a variety of fast jet and warbird types, but the North American T-28 obviously gives him a buzz.

Sitting down in the cockpit you are immediately impressed by its roominess and superb layout. The seat adjusts up and forwards/ down and aft and the pedals are adjustable forwards/aft, making it comfortable for a large range of pilot sizes. taking a look inside, starting on the left side there is a manual emergency hydraulic pump, beside it on the console is the cockpit air control. Canopy control lever and an air nozzle is on the wall above it. In front of the air control is the fuel selector, and that completes the console. Next (on the left) we have the throttle quadrant containing: Throttle, pitch, mixture, blower, flaps, and carburettor air controls. Forward of this is a small panel with the cowlflap switch and hydraulic pressure gauge. The left side panel has the ADF indicator and speedbrake warning light. On the left side of the main instrument panel is the gear handle and gear position indicators. The rest of the panel is set up with a basic 'T' instrumentation and the engine gauges to the left. On the right we find the tailhook handle and parking-brake handle. The right sidepanel contains the LORAN indicator, the ignition switch, and the oxygen flow indicator. On the right console from the front we find the starter and primer buttons, pre-oil switch, electrical panel with cockpit lighting controls, dual NAVCOMS, ADF, LORAN, transponder, map case and oxygen controls.

Phew! Now let's have a look at the aircraft from the outside. Standing on the wing you can check oil and hydraulic quantity by opening two small hatches just in front of the windscreen. Thereafter you walk out on the wings, open the hatches and fuelcaps, check the fuel visually, reinstall the fuel caps and close the hatches. *Never trust the fuel gauges!* Getting back down on the tarmac using the aforementioned steps, bend down under the belly and open the belly hatch. Standing up inside the belly turn on the light (provided due to the genius of a *North American Aviation* design engineer), and boy is there a lot to look at! Flight control belcranks, pushrods, balances and wires. Flap, speedbrake, and canopy actuators. Canopy emergency nitrogen bottle, oxygen bottle and pressure indicator. I usually check the battery connector from here. There is a

lot of electrical wiring, hydraulic and pneumatic plumbing and generous space for your (or your wife's) luggage. Remember to turn off the light, and close the hatch. Next check the speedbrake which is right in front of the belly hatch.

The flaps, ailerons, and tabs are checked, fuel is drained to check for water, tyres, brakes, leg extension, gear doors and wheel-wells for condition and leaks. Remove the downlock safety pin. Moving on to the left forward side of the fuselage you notice an inlet in the left wingroot. This is for the cockpit heater, but this system has been removed in our aircraft. On the bottom side I drain some more fuel, and then check the nosegear well. All the small hatches must be closed, and the stepladder must be stowed. No, you didn't misread! There is actually a stepladder there to let the mechanics climb up and work on the engine accessory section after having removed the canvas in the top of the wheelwell and part of the firewall. There is also a light installed here for your convenience, all provided for by that same nifty engineer at N.A.A. Check the shimmy damper for proper indication, and the rest of the gear. The cowling is checked for being properly secured, and the big Wright Cyclone engine for leaks, condition, and last but not least foreign objects. I have found coke bottles, candy-paper, cups and a lot of other things in there after static displays! The right wing and forward fuelage is similar to the left with the exception of the pitot tube and there is a small fuel reservoir in the wheelwell. At the tail section we check the horizontal stabilizer, elevator and tabs. There are two bolts on the inside of the balancing horns on the elevator that have to be checked specially, as they can jam your elevator if they're loose. As they say "A small thing like that can ruin your day". Next we check the rudder and notice that there is a bungee that returns it to neutral after being deflected. This is part of the artificial feel system on the T-28. Ailerons, elevators and bobweight, bungees etc! Some of these are ground adjustable. We check the rudder trimtab and the fin. The tailhook is next, and we see that the uplock is in place and so is the centreing cam. It uses pneumatic pressure for extension and hydraulic for retraction. On the left side of the fuselage there is a battery access door, but as I mentioned earlier I check the battery from the inside.

Well, we are back at the left trailing edge of the wing, so let's climb up into the cockpit again and go for a ride!

You strap on the parachute and climb into the seat. Secure the lapbelt and inertia-reel shoulder straps. Next is the bonedome, with the oxygen mask. I always fly this A/C with oxygen for two reasons; - exhaust fumes in the cockpit and in case of an inflight fire. Normally there are no fumes in the cockpit, but under certain conditions you can detect them. If I have an inflight fire I don't want to be incapacitated so that I can't extinguish it or bail out. Next are the gloves, so my hands won't slip on the controls and if I get that fire I can use them. Not much use having oxygen, if all you can do is to look at your hands which remind you of shish-kebab.' So get that flightsuit and the rest of the gear to protect yourself and thereby your aircraft.

Enough of pointing fingers. All strapped in and ready to go, pre-start list completed. We call *CLEAR* as we check the prop area, and get a thumbs up from the crew chief. Push the starter button, count at least eight blades, push the primer and switch on the ignition. As the engine fires regularly on primer, bring the mixture slowly into the *rich* position and when the engine slows down release the primer. Check oil pressure rising within 10 seconds and to minimum 40 psi in 20 seconds. I warm up the engine at about 1200 RPM, with the cowl flaps full open to get a nice gradual warm-up. I check the hydraulics by exercising the flap and speedbrake. With pre-taxi items completed we start taxi. The Trojan needs little power to taxi, about 700-900 RPM, except for sharp turns where you have to use rudder, brake and a little power. Taxi is straight forward using rudder only on straights and differential braking in the turns. Even though this aircraft has a nose wheel it doesn't have nosewheel steering. At the holding position we check the engine instruments for the correct values, and perform the runup. Cycle the prop twice at 1600 revs, looking for a 400 drop in RPM, then increase to field barometric for the powercheck, increase to 2300 revs and check the mags for a max drop of 75. I test the blower every now and then, but not on every flight. Special for the C model is a vibration range between 1900-2200 RPM on the ground. At 1800, oil pressure should be minimum 65psi, and idle around 750. Finishing the before take-off list, I do a last check of the flight controls, and then close the canopy using the control lever in the desired position and activating it with a push button on the lever. If for some reason the other crewmember wants to stop canopy operation, simply push the emergency stopswitch on top of the glareshield in either seat.

Lining up I look for 5° right on the rudder trim and neutral on elevator and aileron. When cleared to go I run up to 30" MAP on the brakes, close the cowl flaps to ¼ open, and release the brakes. Smoothly advance the throttle to 48" MAP (52.5 max at sea level) Acceleration is good, and as the nose wheel lifts off you will have to push some more right rudder. It requires a lot more rudder than a Harvard T-6, but less than a Mustang or Corsair. Let it fly off at around 90 knots IAS and select gear *up* when definitely airborne, and when there isn't enough runway left to make a safe wheels down landing in case of an engine failure. I let it accelerate up to a climb speed of about 140 knots IAS and reduce to 36" MAP and 2 400 RPM for climb power at a safe altitude, (taking into consideration the terrain conditions). The Trojan has a very good climb. In fact it is capable of more than 4 000 feet/min. That's in the same ballpark as most World War II fighters. Reaching 5 000 feet I level off, set 30"MAP and 2 000 RPM, and trim the A/C. It is easy to trim and settles down at about 185 knots IAS. Stability is good, and thanks to a 5° down tilt on the thrustline you don't have much destabilizing effects with power changes. The controls are nice, crisp and responsive with a good harmony. They heavy up a little with speed, being stiff at 300+ knots IAS, but not as heavy as the P-51. Exploring the flight characteristics of the Trojan you have to know that there are some restrictions on G loadings versus roll. You can only pull 2/3 of the normal G load if you have any rolling motion. On a smooth day the normal limit is 5G's at 170 knots IAS and 6 G's at 210 knots IAS (the dynamic stall line). In moderate turbulence the limit is 4 G's. With this in mind let's do some manoeuvering. I use climb power for aerobatics and you don't need any more. Primary, aileron and barrel-rolls are a blast. The roll-rate is better than the P-51. Over the tops can be flown from 170 knots IAS and up if you are careful. It seems like the Trojan hits the wall at about 250 knots IAS in a moderate dive, but if you dive at 30° or more it builds up to 300+ fairly quickly. Don't use the speedbrake in the dive recovery as deploying this above 250 knots IAS gives a pitchup. This combined with a pullout can increase your G load by 2-3 G's, thus overstressing the aircraft. It is fascinating to watch the contrails coming off the prop at slow speed which is quite marked in the Trojan compared with the Harvard or Mustang, but it can be a disturbing factor on top of a loop. Talking about slow speed, the stall with power on is the normal left wing dropping and nose pitching down. The controls are effective down to the stall and recovery is straight forward. One thing to remember is that there is little or no warning of the power on stall. A big *WARNING*: pulling the Trojan into a highspeed stall and letting it snap, can easily lead to a structural failure and even the departure of your horizontal stabilizer. (A big thing like that can also ruin your day).

This aircraft is a lot more agile than most people think. Once you have flown it for a while, and get comfortable in it, it will downright impress you. Sometimes I start my display with a roll on takeoff and then straight into an Immelman, although I prefer to take off before the slot time and start with a vertical rolling manoeuver. In fact, it will do a double

WARBIRDS WORLDWIDE

TERRIFIC TROJAN

Immelman, but that is pushing it. With this agility it's quite fun to be jumped by a fighter, and force him into a slowspeed (250 knots IAS or less) dogfight. Then you can give him a real hard time, with turns, scissors and you name it. If you get into a dogfight with a P-51 you can outfly him in most respects other than speed and acceleration in a dive. In fact you can dogfight with just about any WW2 fighter on equal terms below 300 knots IAS.

Heading back for the field I aim to be on downwind at 120 knots IAS with checks completed and gear down selecting full increase RPM, and flaps down when turning base. Flying base at 100 knots IAS and final about 85 knots IAS. The landing is straight forward, observing the nose wheel touchdown speed for any appreciable crosswind. There is no steering on the nosewheel as mentioned earlier, so you have to use the brakes when the rudder effectiveness diminishes. This is at a higher speed than you would think, in high crosswinds. It isn't any big problem though.

Taxiing in and shutting down, I make sure of scavenging the engine and check the cylinder. head temperature. below 150deg.C. Once stopped there isnt any cooling airflow over the engine, and the retained heat may cause damage to components.

Well, that's a few words about flying the *Scandinavian Historic Flight's* T-28, and to finish off I would like to answer a few of the questions I get about the aircraft:
- NO, it's not a Harvard, Hellcat or Thunderbolt.
- NO, it didn't fly in World War II.
- Yes, it has been used by the USN.
- Yes, it is expensive to fly.
- Yes, it is more difficult to fly and operate than a Harvard, except for landing.
- Yes, it is heck a lot fun to fly.

WW Rolf Meum

The North American T-28 is a tandem two seat basic and shipboard trainer or Light Tactical and counter-Insurgency aircraft. The aircraft is larger than you might think; Wingspan is 40ft 1in, length 32ft 10in and is 12ft 8in in height. Empty the aircraft weighs just over 5100lbs. For those T-28 (and T-6) fans the *North American Trainer Association (NATA)* is highly recommended. Contact Kathy Stonich at: *NATA*, 25801 N.E. Hinness Rd.,Brush Prarie, WA 98606 Tel: (206) 256 0066

Mike Barrow zinc chromating Joe Scogna's P-51D at Classic Air Services, Cape May, New Jersey (Jeff Ethell)

Though *Classic Air Services* at Cape May Airport, New Jersey is a new company, its beginnings reach far back into the warbird movement. In 1964, when P-51s were still fast executive transportation with slick civilian paint jobs, Jack Shaver got involved with his first Mustang. By the early 1970s he was known as one of the experts on the fighter, a reputation that has followed him for more than 25 years.

When east coast Mustang owner/pilot Jim Beasley had his regular maintenance done by Jack the two hatched the idea of a full-service warbird operation. In 1988 *Classic Air Services* moved into the large ex-World War II U.S Navy hangars at Cape May. These massive hangars are ideal for warbird maintenance and operations, though Jack and Jim are not sure if they are going to remain in these facilities or build new ones on the field.

Cape May sits on the southern tip of New Jersey due east of Washington, DC so it is ideally located on the middle east coast, very close to the major cities by airline. The four 5000 foot wartime runways are intact, making it about the easiest airfield to land on, regardless of wind. An ILS approach is in use and there are plans to extend some of the runways to 7,000 feet for high performance jet operations.

Jeff Ethell reports on this warbird facility, ideally located at Cape May, New Jersey, which already has a reputation for excellence without a sting in the tail

Classic Air is offering the full range of general maintenance, repair and ground-up restoration of warbirds under Jack's leadership, with Mike Barrow as Chief Mechanic and Norman Wells as one of two part time assistants. Jack intends to maintain his reputation for being one of the best at keeping warbirds safe while not running up an enormous bill. Owners around the world find they are often hit with massive bills just because they show up with a warbird. Certainly the vintage aviation scene is one that generates high cost but owners are often the victims of goldplating. As *Classic Air's* operation grows more people will be put on the payroll until a well rounded staff of good mechanics are on duty full time.

Last winter two P-51Ds were under full restoration. Joe Scogna's *Baby Duck* was completely gutted after several of us flew it in 1988. Though we knew it was a weary aircraft (see "Mustang Summer," Warbirds Worldwide Number 8), I don't think we knew just how weary. When Dwight Thorn pulled the engine apart for a top overhaul, he found everything loose from the nosecase back. Though the Dominican Air Force had listed it with 65 hours running time, Dwight said it had 500 if it had a day. That's a good idea of just how hard the Dominicans ran their engines . . . after all, these were military aircraft flown by pilots who used maximum power a great deal. Joe authorized Dwight to do a total overhaul to zero time specs and then instructed Jack to make the airframe match.

The wiring had been baked in the hot Caribbean climate until the rubber insulation became brittle and fell off. The hydraulics were a weeping mess on almost every flight. Both systems were replaced with brand new material. The cockpit was disassembled totally with everything being stripped and repainted and a new instrument panel was designed and installed along with a new radio stack. About the only thing not replaced was Dan Calderale's beautiful World War II paint job.

The other Mustang going together is Jack's personal airplane which he has been working on for several years. Every last nut, bolt and rivet has been worked over and Jack has been rewiring to his detailed specs so that the inside has the systems and look of a jet. When this machine flies it will be one of the finest restorations around.

Classic Air Services

CONTINUED ON PAGE 34

FROM THE PUBLISHERS OF THE WORLD'S NUMBER ONE WARBIRD PUBLICATION
NEWS OF TWO NEW TITLES

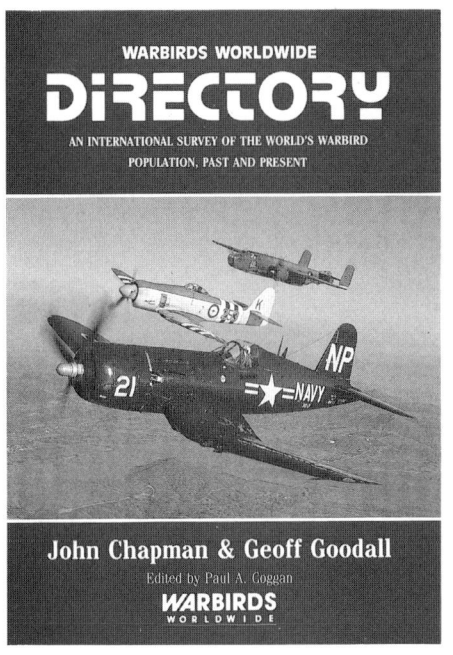

■ WARBIRDS WORLDWIDE DIRECTORY
A5 · SOFTBACK · 320 PAGES

Aviation Historians John Chapman and Geoff Goodall present an International survey of the world's warbird population past and present with this first edition of the Warbirds Worldwide Directory. Softback, with 320 pages in A5 format this book is brimming with data, updated to September 1989 and is the results of years of research by the authors. Types listed are: Lancaster, Lincoln, P-39, P-63, B-17, B-29, Beaufighter, Beaufort, Blenheim, Boilingbroke, P-40, Helldiver, Liberator, Mosquito, Bolo, Dauntless, Dragon, Havoc, Invader, Skyraider, Firefly, Fw190, TBM Avenger, Bearcat, Hellcat, Tigercat, Wildcat, Iraqi and Sea Fury, Hurricane, Tempest, Typhoon, He111, P-38, B-26, Bf109/Buchon, Me262, B-25, P-51, F-82, P-47, Sunderland, Spitfire, Corsair, Jet Provost, Vampire, Venom, Sea Vixen, Skyhawk, Lightning, Gnat/Ajeet, Meteor, Cougar, Panther, Seahawk, Hunter, T-33, F-104, MiG-15, MiG-17, MiG-19, MiG-21, and F-86/CL-13B.

AVAILABLE NOW!

Prices inclusive of airmail postage (where applicable) are: U.K.: £12.95, Europe and Scandinavia £14.50, U.S.A. $24.95, Canada $CAN 28.00, Australia $AUS $28.00, New Zealand NZ$34.00.

■ WARBIRDS WORLDWIDE SPECIAL EDITION
BOEING B-17 FLYING FORTRESS
A4 · SOFTBACK · 48 PAGES

From stunning images of the B-17s caught in action during the filming of Memphis Belle to the article on flying the *Fort* by Jeff Ethell this *WARBIRDS WORLDWIDE SPECIAL EDITION* pays tribute to this much loved American bomber. Special features include: **Flying the B-17** by warbird pilot and writer Jeff Ethell, **Black Jack's Last Mission** by Steve Birdsall, **Baby Comes Home** by Robert S. De Groat – the long and sometimes controversial rebuild of B-17G *Shoo Shoo Shoo Baby*, **Memphis Belle II** by Clive Denney – Clive chronicles the story of the return to the USA of David Tallichet's B-17 which played the starring role in the film. **Sentimental Journey** – Alan Gruening covers the rebuild and operation of the *Confederate Air Force Arizona Wing's* B-17, and Alain Wadsworth takes a mainly illustrated look at the **B-17s** of the *French Institut Geographique National*.

AVAILABLE NOW!

Prices inclusive of Airmail Postage (where applicable) are: UK: £4.50, Europe and Scandinavia £5.00, U.S.A. $8.95, Australia AUS$10.00, New Zealand NZ$10.00, Canada $CAN10.00.

We accept VISA, ACCESS, MASTERCARD AND AMERICAN EXPRESS.
Why not Fax or Telephone your order through to our 24 hour lines below?
WARBIRDS WORLDWIDE Five White Hart Chambers, 16 White Hart Street, Mansfield, Notts NG18 1DG
Telephone (0623) 24288 or Fax (0623) 22659 (24 hours)

Continued from Page 32

Top: *A visiting T-28B, NX514FR is seen in the massive hangar at* Classic Air Services *Cape May base.* **Lower**: *Jim Beasley's SNJ will form a part of the fleet of aircraft for the planned warbird school where already experienced pilots will be able to improve their skills. Those interested in the services offered by* Classic Air *should contact them on (609) 889 0300 or at Cape May County Airport, Cape May, NJ 08242, U.S.A.*

In addition to the shop, Jim and Jack are planning to create a complete warbird training school using a T-6, Jim's dual control P-51D (ex-Dominican FAD 1900), a T-28 and possibly a T-33. Pilots can come and stay in Cape May (one of the finest beach resort areas on the east coast) while going through their training. Since the airport is located on a very flat area, transitioning into a high performance aircraft will be much less hazardous than in a crowded urban area. In the evenings one can relax on the beach and eat in some truly outstanding resturants. Bringing the family along would be a natural.

With the advent of standardized formation flying from *EAA's Warbirds of America* in the T-34, T-6, P-51 and T-28, Jim is planning to hold long formation training weekends for owners and pilots. This would include formation take-offs and landings which can only be practiced on wide runways. Biennial Flight Reviews will be available as well. Jim has made it clear, though, that the programme is for transition into a specific type, not learning to fly. It would be very helpful if someone had a fair amount of tailwheel time logged. In addition to flying, systems and manuals will have to be mastered, emergencies simulated, such as landing on one wheel and holding the wing up to simulate this less than rare problem.

If someone wants to leave his aircraft, all he has to do is hop on the commuter airline and within less than an hour be at any one of several international airports. If the owner needs to get out at a time not served by the airline, an Aerostar sits on the line ready to do the job.

Classic Air Services will fill a major gap on the east coast for warbird maintenance and transition, offering a vacation for the lone warbird driver or the whole family. Anyone interested can enquire at: Classic Air Services, Hangar 1, Cape May County Airport, Rio Grande, New Jersey 80242, phone 609 889-0300.

WW Jeffrey L. Ethell

WARBIRDS WORLDWIDE

Jack Shaver's North American Mustang rebuild which will be one of the finest around once complete (Jeff Ethell).

JETTOPICS

Charles Parnell leads up the jet scene with a report on the latest happenings.

It must be that spring is in the air, because jet activity is certainly on the increase. There have been several changes in the Vampire/Venom camp, with Ernie Saviano of Portage, Wisconsin, heading the list by purchasing one of each! The reliable and aesthetically pleasing Yugoslav built Soko *Galeb* Two place jet trainer is making a good impression with its low maintenance, 'reasonable' (for a jet) fuel consumption and more than adequate performance. Prices have escalated rapidly on this Rolls-Royce Viper powered aircraft, as more and more prospective owners look to the jet era for a source of flying enjoyment.

Another trainer that could surface soon is the Czechoslovakian L-29 *Delfin* jet trainer. Due to import restrictions these aircraft are prohibited from entry into the United States, but current conditions, along with a recent loosening of the laws may find this trend reversed and the type could soon be airborne. Any time a vintage jet takes to the air history is shared with a wider, more educated audience.

The T-33 continues to be a firm favourite. At Fort Wayne, Indiana, John Dilley is heading down the final stretch with a complete restoration of his Canadair Silver Star (c/n98, ex RCAF21098) N99184, whilst in the same State Rick Ropkey, and Hampshire based Lou Antonacci should both have their T-33s flying this year.

Fast becoming one of the largest jet restoration facilities in the United States is *Sierra Hotel Inc.*, of Dallas, Texas. President David Strait reports that they have recently moved to a large (20,000 square feet) hangar, and have no less than five Folland Gnats, an A-4B Skyhawk, a Lockheed T-33A a Paris jet

Top: Combat Jet's *latest acquisition, a MiG-21!* Centre & Below: *Currently there are only two Douglas A-4 Skyhawks being operated by civilans. These are Sierra Hotel's Australian Navy marked aircraft and Combat Jet's U.S. Navy marked aircraft in VA-76 markings. Lower Photo by* Joe Cupido.

JETTOPICS

and a Soko *Galeb* on charge. Their plan is to form a four ship jet formation team with the Gnats, which to our knowledge will be the first civilian four ship jet team in history. The fact that four of the five aircraft are ex Red Arrows aircraft will certainly return these celebrated machines to their rightful position in the air – and might even make it easier for their new American pilots, since the aircraft surely know how to fly in formation! With some luck, the team hope to be fully operational by mid-1990. As one might expect operating costs for this venture will be significant, and a soon to be announced major sponsor is at hand. Strait's A-4B Skyhawk has been doing some commercial work recently most notable being a *NARCO* advertisement. This rare bird is painted in Australian Navy markings and has logged over 60 hours since it first flew in 1988. Dave says it is a dream to fly, but warns the uninitiated that careless inattention will be rewarded in a deadly manner. Another project scheduled for this spring is the restoration (to static condition) of an A-7 Corsair II. Past accomplishments include T-33, Vampire and Grumman Cougar jets. At *Sierra Hotel* training is as important as maintenance and the company

Heading photograph: shows BAC Strikemaster 305 of the Singapore Air Defence Force during test flying with BAC (British Aerospace). **Centre:** *the fuselage of Strikemaster 310.*

recently completed a programme with their local FAA office to perform check rides in jet warbird aircraft: an accomplishment that will, no doubt, stand them in good stead as time passes. *Sierra Hotel*, another major player in the jet arena.

From Australia, Peter Anderson reports that Sydney based IAP have imported several BAC Strikemaster jets from Singapore. Of the fifteen aircraft, Numbers 304, 305, 308, (Mk.82), 310, 311, 312, 314, 315, 322, 323, 327, 328, 329, 330 and 331 only three have been imported to Australia. The remaining aircraft were shipped directly to the United States and stored in Los Angeles but will now be moved on to Australia for overhaul and sale. Enough spares have been acquired to maintain the aircraft for at least ten years under a military training regime of 400 hours per year, but with the aircrafts total times only averaging 4243 hours this spares holding will provide support for at least double that time. Ten of the aircraft are currently allocated to a foreign customer for use

WARBIRDS
WORLDWIDE

Continued on Page 44

37

My introduction to the MiG-21F was *awesome*. As a young second Lieutenant with some 300 hours of jet experience in Fouga Magister trainers, the Mach 2 supersonic interceptor was a giant leap from turning in the sky at speeds below 300 knots. The all-new supersonic world was ours in the silver dart like performance pack that was named the *MiG-21F*.

The Finnish Air Force purchased its first MiG-21Fs in 1963. Though many western and Finnish aviation sources identified them incorrectly as Dash 12s they were in fact Dash 13s. My introduction to the aircraft was in 1970, and the MiG was no longer the secret super weapon it used to be in the early sixties. But to us it was the best fighter in our service. Though lacking radar weaponry and long range, it was a hot performer in climb, speed and turning. The new dogfighters were still to come.

Learning to fly the MiG-21 started with groundschool. This included basic

Test and warbird pilot **Jyrki Laukkanen** takes us on the trip of a lifetime in a MiGF-21F of the Finnish Air Force.

system, primary system and booster systems, each of which is supplied by its own pump. In an emergency there is an electrically driven standby pump to supply the booster system.

Ailerons are hydraulically powered by the booster system with direct mechanical linkage. The rudder is all manual (push-pull rods) and has no power system. The primary hydraulic system also operates the gear, flaps, airbrakes, intake cone and jet nozzle. The gear can be lowered by an emergency air system.

During ground school we became familiar with the Soviet style of abbreviations of different systems and units like ARK, RV-UM, SRD, AGD, KSI, DUS, RAU, ARU and APA. These are simply the first

ing when trying to deploy the drag chute. Practice makes a master and normally we had no real difficulty in operating the switches and knobs. Though we learned to find them with our eyes closed, it was better to look at the switch before using it!

The pilot's ejection seat is designed to work safely at minimum altitude of 130 metres (430 ft) while in level flight. The canopy is designed to eject with the seat to protect the pilot from the high dynamic air pressures during high speed ejection. There is a limit of 95 cm (38 in) on the length of the pilot's back. Taller pilots are in danger of their head being squeezed between the headrest and the attached canopy. Later model MiG-21s, starting from the PFM, have fixed separate windscreen and normal type sideways opening canopy that is jettisoned before the seat is fired, and has no such length restriction for pilots.

The external walk-around is fairly normal. Climbing to the cockpit requires

The Awesome MiG-21

aerodynamics and a study of supersonic flight, aircraft systems, flight manual and flying procedures.

The MiG's aerodynamics are pretty straightforward; the tailed delta configuration with a thin 4.3% airfoil section gives good wave drag qualities at supersonic speeds. It also allows the use of normal trailing edge flaps to reduce the high takeoff and landing speeds associated with the small wing area.

A good looking aircraft normally has good flying qualities. And so does the 21F. Both static and dynamic stability are naturally so good that no damping systems are required for any axis. This is unusual for Mach 2 fighters.

Control systems are as simple as possible and typical for that time period. The stabiliser is powered by a dual hydraulic

letters of the systems names in Russian, but we quickly adapted these as the actual names.

Cockpit ergonomics are a far cry from today's pilot friendly environment. Over one hundred switches and knobs are spread all over the cockpit in an untidy manner. Though there were some grouping of these to help the pilot, it took some time to pass the cockpit switch check in which you had to find each item with eyes closed. Even the main flying instruments are spread in the panel in areas where there is room for them. The T-type arrangement was still to be recognised by the MiG design bureau in the fifties.

Even as we studied the cockpit, some pilots still managed to arm the 30mm gun with the drag chute button and then drop the missile launch rails during land-

separate ladders, because there are no built in steps. The seat type parachute is installed in the ejection seat and has a common harness with one central lock to work with.

Before start-up procedures require a lot of items to be checked and you have be careful not to leave any switch in the wrong position. The start-up itself is not too complicated once you have got through the checklist items. It is "automatic", Russian style. They used to call automatic almost any function where you need not perform the task manually. The gear retraction is automatic as it happens by selecting the gear handle into the up position and the pilot does not need to crank the gear up.

To start up the engine with an electric starter you normally need external

39

power. There is no sense in drying up the battery doing that. The special generator equipped truck is called APA. Throttle to idle position, push the start-button and the engine will start up. All you need to do is to check that the START-light comes on and then watch RPM spool up to about to 35%, oil and hydraulic pressure rising and check that the jet pipe temperature remains within limits, which is never a problem if the throttle is at the idle position. Once the start-up sequence is complete the indication light goes out and your powerplant is ready to go.

Before taxi checks just call for switching the necessary systems on and control checks for primary flight controls, flaps and airbrake. Steering during taxiing is accomplished by differential braking using air pressure from reservoirs. Squeezing the handle in the control stick gives you proportional brake power which is divided to the main wheel brakes by rudder pedals; the same way as in my vintage 1930s Gloster Gauntlet. Turning during taxi is a difficult business due to the small main wheels and long fuselage. Frequently on icy taxiways you could not turn or stop your supersonic jet and would end up in the snowbank!

Once on the runway heading, the MiG goes as straight as a train during the take-off run. Before releasing the brakes you have to do just four last check-list items: nose wheel brake *ON*, gear handle ground safety switch *OFF*, heading *SET* and *CHECKED* and pitot heat *ON*. Check the power indicators and let it go.

Take off with a clean aircraft can normally be undertaken on military power only, without the afterburner. The aircraft becomes airborne after a 4000 ft run. Using full reheat the MiG-21F leaps into the air in less than 3000 ft. Take off is easy to fly as the aircraft maintains the heading almost by itself. The nosewheel is lifted off the runway at 250 km/h IAS (135 knots IAS) and the rotation initiated at 320 km/h IAS (175 knots IAS). The pitch control is nice at rotation, and the aircraft lifts off almost by itself after rotation to 10 degrees nose up attitude.

There is hardly any pitch trim change while retracting the gear and flaps. The wheels have automatic braking while retracting. If you are in a hurry to get to the stratosphere, with full afterburner the MiG accelerates very nicely in almost level flight to its best climb speed of 1000 km/h TAS (540 knots IAS). Once you've got that indicated, pull the nose up and maintain it. You will be climbing through Flight Level 350 in less than two minutes from brake release. You still have the nose up so much that for levelling off it is more convenient to roll inverted and pull the nose to the horizon and then half roll again, instead of pushing negative G and hanging against the canopy. (Editors note: and there are those that won't accept the MiGs as Warbirds!)

This type of climb sure gets you to altitude in a hurry. But for more conservative flying 93% of N1 (low pressure compressor rpm) is recommended to save engine and fuel for climbing to the practice area. Climb initially at 700 km/h IAS (380 knots IAS) until reaching 930 km/h TAS (500 knots TAS). This will give you the most economic climb. The airspeed indicator has needles for both Indicated Air Speed (IAS) and True Air Speed (TAS).

If there is a need to climb above 35 000 feet, you'd better use your basic knowledge of supersonic fighter performance. The MiG-21F has a movable cone supersonic intake system and variable area jet nozzle that gives you more thrust the faster you go. So if you want to go to altitudes in the region of 60 000 feet, you'd better accelerate to something like Mach 1.8 at 33 000 feet and then start to climb. With the afterburner in the kerosene is going too fast for pleasurable warbirds flying, but you go to an altitude at a speed that you cannot experience in anything other than a supersonic jet. After this spectacular climb be prepared to return to base because you have just burned most of your fuel.

The MiG-21F handles nicely at high supersonic speeds even without any autostabilization. Going through the sound barrier is noticeable only by checking your pitot-static instruments as the shock wave passes through the pitot tube. As the pilot is sitting in the middle of the air intake channel to the engine, there is considerable noise at higher supersonic speeds, like Mach 2.

Whilst descending from altitude you will be suprised how easy it is to maintain supersonic speeds with idle power keeping the nose down only 10 degrees. This is due to good aerodynamics and partly to high idle rpm at altitudes. Coming down from altitude with idle power burns almost no fuel at all. You will burn more kerosene in a single go around than you need during a descent from a distance of 100 nautical miles. So plan your approach carefully. Coming down clean is almost a free ride. Going up dirty will cost you a great deal in terms of endurance.

There is no sense in maintaining an IAS below 500 km/h (270 knots) at any time as the induced drag (due to the low aspect ratio wing) begins to punish you. You begin to feel like hanging in the sky nose up and going nowhere! Good overall speed for any cruising is some 700 km/h IAS (380 knots IAS). There is no angle of attack indicator in these early model MiGs, so you have to hang on the IAS.

To further reduce your airspeed requires more and more pull on the stick, as the MiG-21F is very stable statically. No relaxed stability with Fly-By-Wire-controls to limit you. The aircraft handles exceptionally well at low speeds. There is enough aerodynamic control to the point where the stick is almost all the way back. There is ample warning of slow speeds but they will not hurt you, unless you are at low altitude. The sink rate is spectacular and you had better practice this kind of slow flying in excess of 20 000 ft at idle power.

You can safely slow down to some 200 km/h IAS (110 knots IAS) with the stick all the way back and still have lateral control of the aircraft, though it

slowly rolls to both sides. Your landing speed will be 320 km/h IAS (175 knots IAS) over the threshold, so there is a lot of margin in slow speed handling on final approach. Recovery from a stall is simple. Put the nose down and let it get airspeed back to good numbers like 500 km/h (270 knots) and you will be flying again.

The aircraft is very spin resistant and it is hardly possible to get it into any post stall gyration without positively applying full rudder and aft stick at high angle of attack; this makes no sense anyway. We never spun the aircraft accidentally or intentionally either. There is no need for this kind of manoeuvre and so no need to practice it.

The engine spool-up time is longer than your patience. So as I said, don't practice these things at circuit heights. On occasion pulling the throttle back and trying to get it back to a higher setting has made me sense the engine has failed. So don't do that at low altitude, because it may increase your heart rate significantly!

Visibility from the cockpit is fairly good and much better than in the later models. To turn the sky around at low and medium altitudes you don't need to use the afterburner. Just keep the speed over 700 km/h IAS.

The MiG-21F's static stability is good in any axis. The dihedral effect, or roll due to sideslip, of the swept delta wing is especially effective. Release the stick, push half rudder and you roll the aircraft 360 degrees very nicely in a few seconds.

Flight controls are nice and effective but not oversensitive. The pitch control operates the all moving tailplane with a double hydraulic system with no mechanical back-up. Control forces come from the spring so they are convenient. There is a q-feel system called ARU, that changes the pitch control gearing proportionally with speed and altitude. At low altitudes and high IAS the stabiliser moves less. At low IAS and high altitude it moves more with the total stick movement being the same. This prevents overcontrolling at high IAS and still gives you enough elevator power at high altitudes and slow airspeeds.

The aileron control is very effective at normal flying speeds before you've got the stick to the other side, the aircraft has rolled 360 degrees with your head on the other side of the canopy. The ailerons can be operated manually in case of booster system failure. With the mechanical control, however, the stick force is really heavy and it is for emergency use only. A better way to control the aircraft in a roll in this case is to use the rudder and the very positive dihedral effect.

The manual rudder control forces are light at low airspeeds but become pretty heavy at high speeds.

'So you can come into land even with almost zero fuel. The only limit is to keep the engine running! For safety reasons zero fuel reserves make no sense'

Some western sources say that there will be almost one third of unusable fuel because of the centre of gravity restrictions. We have not heard about that. Indeed the fuel consumption changes your CG position a little while flying, but it stays within limits all the time. So you can come in to land even with almost zero fuel. The only limit is to keep the engine running! For safety reasons zero fuel reserves make no sense. If you have 500 litres on landing you can still safely make one go-around if required. There is a red warning light illuminated when 500 litres of fuel remain.

Always plan your approach in good time, because once you are getting low on fuel, it seems to be burning really fast. The airbrakes are effective in slowing down your airspeed for gear extension. Once the speed is below 600 km/h IAS (325 knots IAS) you can lower the gear. There is a minimum nose up trim change that will be mostly countered by selecting the flaps to the down position, which is the same for take-off and landing. This makes the go-around easier.

Once you have the landing configuration selected, added power is required to maintain airspeed. The pattern has to be large enough to allow shallow turning so as not to overfly the final and lose all your airspeed in a tight low speed turn. The induced drag of the low aspect ratio delta wing is very impressive. If you get too slow and low there is not enough thrust (even with the afterburner) to fly you out without losing height while reducing the angle of attack.

Once on the base leg, slow down to 400 km/h IAS (215 knots IAS), check again that the gear is down with three greens and hydraulic and air pressures are OK. Check that the *STABILIZER* light is illuminated to advise you that the ARU is in the correct gearing to give you full stabilator movement for landing.

Flying a five mile long final will give you enough time to adjust to correct landing speed. Once used to the aircraft handling and performance, a shorter pattern can be flown. But be careful not to slow down too much and not to have excessive speed either. During long finals you can gradually slow down your airspeed from 400 km/h (215 knots) to some 340-320 km/h IAS (185-175 knots IAS) over the threshold. Once over the runway start the flare quite normally whilst reducing power to idle. If done properly, the MiG-21 touches down nicely on the main wheels. With a little practice you can really smooth it out!

On final approach over the threshold there is little if any view of the runway due to the high angle of attack. So the height during flare must be viewed from either side. This is no real problem. However even moderate rain will result in no forward view through the flat windscreen. During normal straight finals you cannot see the runway at all. The only way is to make a tight turning approach, but to do that you have to be well experienced with the aircraft. So it is better to save the landings for days when there are no rainshowers around!

The aircraft will touch down at about 260-280 km/h IAS (140-150 knots). If there is enough runway available, you can maintain 10 degrees nose up attitude and let the airspeed slow down to some 200 km/h (110 knots) before the nose will come down by itself. The wheelbrakes are effective and even the nosewheel has a brake. Due to the high landing speed you need some 2 000 metres (6 500 ft) for this kind of smooth landing. There are is no difficulty in maintaining your heading during the ground run.

If you need to use less runway, it is advisable to touch down at 260 km/h (140 knots), deploy the drag chute, let the nose down right away and start braking immediately. By doing this you can squeeze the landing run easily into 1 000 metres (3 500 ft).

The MiG-21F is a straightforward aircraft to fly with pretty good performance. With adequate training and experience warbird pilots should have no difficulty in operating it. The small delta wing means it is not a STOL aircraft and the runway requirement for safe operation is 7 000 ft. Its systems are simple but need technical know-how. With proper maintenance it is also a reliable aircraft. Though the modern fighter technology has made the MiG-21F in many aspects obsolete as an effective fighting machine, it will be one of those famous fighters in world's aviation history and has earned its place among the warbirds.

WW Jyrki Laukkanen

Number Thirteen

In Warbirds Worldwide Number Thirteen the **Editor** reports from the *Classic Jet Aircraft Association* convention at Tucson, Arizona and on the *New Zealand Warbird* scene with some exclusive interviews. **Thierry Thomassin** visits Chino and details the latest activities there. We complete **Robert Rudhall's** article on the Battle of Britain Memorial Flight, and catch up with the Yak scene. **Philip Warner** begins a major series on the *North American T-6* and we begin our listing of *North American T-28s*

Publication date May 25th – available approximately six weeks later to non-subscribers in the U.S.A.

WARBIRDS WORLDWIDE

The World's Number One Warbird Publication

FAA Repair Station No. 212-23

Covington Aircraft Engines, Inc.

Major Overhauled Engines
Specializing in Pratt & Whitney

R-985-AN1 or 14B
R-1340-AN1
R-1340-S1H1-G

P.O. Box 1344, Municipal Airport
Okmulgee, Okla. 74447, U.S.A.
TEL 918-756-8320
FAX 1-918-756-0923
Telex 3791814

Continued from Page 37

in a training role, but the remainder will be for sale on the warbird market. Anyone interested in the acquisition of any of these aircraft or spares can contact *IAP* direct in Australia on (02) 997 8166 or Fax (02) 997 7631 or at P.O Box 483, Narrabean, Sydney 2101, N.S.W., Australia.

Our own *Combat Jets Flying Museum* in Houston, Texas, (where there is always something exciting going on) has been busy too! In January, the cause of the excitement was the sound of a MiG-21 engine running! That's right... a MiG-21! this aircraft has been undergoing extensive overhaul and modifications nearby and hopes are high that the aircraft will be flying by mid 1990. In keeping with museum mandates the aircraft will be painted in North Vietnamese markings (camouflage) preserving yet another piece of history for aviation devotees to enjoy in its natural element – the sky above!

In WARBIRDS WORLDWIDE THIRTEEN the Editor reports from the second annual *Classic Jet Aircraft Association* **Convention in Tucson, Arizona.**

One of the International Air Parts (IAP) Strikemasters being dismantled in Australia (International Air Parts)

- High Quality Cordex Binder
- Holds Twelve Copies

Binders

In a dark blue with gold blocking on the front cover and spine these high quality binders are available exclusively from the Publishers. Prices are (inclusive of postage and packing, airmail where applicable): UK, Europe & Scandinavia £5.25, USA $16.50, Canada $19.00, Australia, $19.00, New Zealand $19.00.

☆ ☆ *No awkward wires!* ☆ ☆
Superb quality finish to hold our quality journals.

WARBIRDS WORLDWIDE   WARBIRDS WORLDWIDE, FIVE WHITE HART CHAMBERS, 16 WHITE HART STREET, MANSFIELD, NOTTS NG18 1DG, ENGLAND.

THE *Flying Leather Aces* COLLECTION

⊙ R.A.F. Sheepskin Flying Jacket

In true, unfailing Eastman tradition, we proudly present our latest addition to the **Flying Leather Aces** collection. Our reproduction of Leslie Irvin's famous and unmistakeable R.A.F. Sheepskin Flying Jacket. After many months of research we can now produce THE most authentic reproduction available.
£310.50 incl VAT P&P

For this and many other superb WWII reproduction flying jackets, such as A-2, B-3 Luftwaffe styles, write or phone:

EASTMAN
LEATHER CLOTHING
Dept WW, 29 Fore Street, Ivybridge, S.Devon, PL21 9AB
Tel: 0752 896874. Fax: (0752) 690579. Visa & Access by mail or phone.
Write or phone for free full colour brochure and leather samples.
Overseas enquiries please send £3.00

JETTOPICS

SIERRA HOTEL INC. JETS

Sierra Hotel Inc., of Addison, Texas have several jet aircraft including these three Gnats (with two more ready to go) and the increasingly popular Soko Galeb below

WARBIRDS WORLDWIDE

45

CHAMPLIN Fw190 Completed!   Continued from Page 15

Williams came from Germany to work with Dave and Charlie. Visitors to the *Champlin Fighter Museum* are able to view such restorations in progress, as, adjacent to the World War I display hangar there is a viewing area overlooking the restoration hangar and shop where maintenance or restoration of aircraft in the collection is always in progress.

At last, on January 19th, 1990, after numerous test runs of the mighty Juno engine with all the cowlings removed, the Museum's pride and joy was fully cowled and ready for its roll-out. Normally clear, sunny, and comfortably warm (even in January) weather prevails in Arizona. On the morning of the planned runs the Fw190 was greeted with a dense fog! Reminiscent of Germany in 1945, with museum personnel dressed in authentic German World War II uniforms, a photo session was the first order of the day! Then, following final inspections of the engine and cowling, chocks in place, tailwheel strut tied down (despite all efforts to locate necessary parts for the brakes, the crew was unsuccessful, and until brake components can be reproduced from drawings, plans to taxi the aircraft must be postponed) the moment of truth arrived.

With Dave Goss in the cockpit and other crew members standing fire guard, the inertia starter was engaged and the engine came to life before a group of fighter aces assembled for the occasion (the *Champlin Fighter Museum* is the home of the *American Fighter Aces Association*). The year long effort of thousands of man-hours finally reached its climax with the authoritative bark of twelve fuel injected cylinders stirring again after 45 years!

Warbirds Worldwide member Douglas Champlin stands by as Dave Goss prepares to start the Jumo 213E during test runs at Falcon Field (TOP). Under stormy skies more typical of germany than Arizona Dave Goss is at the controls during the engine runs.(LOWER)

Now more precious than ever, the Fw190 will not be flown, even when the brake components eventually complete the project. But, seven days a week, this rare, historic fighter will be on display for more than 50,000 visitors that come to the *Champlin Fighter Museum* every year. **WW Alan Gruening**

In Warbirds Worldwide Number 14 there will be a full length feature article by Alan Gruening and Paul Coggan on the Champlin Fighter Museum.